CRUNCH
TIME.

40% OF ALL FOOD GROWN IS WASTED.

CRUNCH TIME.

ODDBOX

 HarperCollins*Publishers*

CONTENTS.

INTRODUCTION.

SET UP A WASTE- FREE KITCHEN. 14

HELLO.

Back in 2016 we tasted a delicious but ever-so-slightly ugly tomato from a market in Portugal. It struck us that we only ever saw identical-looking fruit and veg in our supermarkets, so we did some digging and learnt that about 40% of all food grown goes to waste.*

Since then, we've made it our mission to rescue fruit and veg considered 'too big', 'too wonky' or 'too many' from farms, which we deliver to doorsteps across the UK.

Along the way, we've built up a wonderful community of food-waste fighters. We've shared recipes and tips online and through our weekly box letters. And we've all learnt how to cook more resourcefully, embracing what we call a grower-led way of eating and enjoying what's available.

So now we want to bring everything we've learnt to your kitchen, helping you make the most of whatever's in your fridge, fruit bowl and cupboard.

With the help of Martyn Odell, food-waste activist and chef, plus our resident chef Camille Aubert, we've developed a collection of flexible recipes that encourage you to cook with the ingredients you already have. We've also captured some of our best tips and tricks which will help you reduce food waste at home.

Oddbox has always been about getting stuck in and trying new things, so this isn't a zero-waste rule book to bash you over the head with. Nor is it a collection of strict, complicated recipes that involve a gazillion ingredients. In fact, it's the opposite of those things – a down-to-earth, wonky-round-the-edges, anything-goes kind of a book, celebrating creative, resourceful cooking.

So dip in. Have a big scoop of whatever takes your fancy. And don't forget to show us how you get on – we'd love to know.

EMILIE & DEEPAK
(ODDBOX FOUNDERS)

* WWF, Driven to Waste Report

MEET THE CHEFS.

HI EVERYONE,

I'm on a mission to cut out household food waste by showing people how to get the most from humble everyday ingredients. I believe the answer to the problem of food waste is as simple as eating the food you buy. As Douglas McMaster says, 'waste is a failure of imagination' – we need to get excited about food and the amazing things we can create from simple ingredients.

I've been cooking for more than 15 years and over the last six years my focus has been on food waste. There's a real problem and the solution is simple: eat!

MARTYN ODELL

(COOKBOOK CONCEPT
AND RECIPE CREATOR)

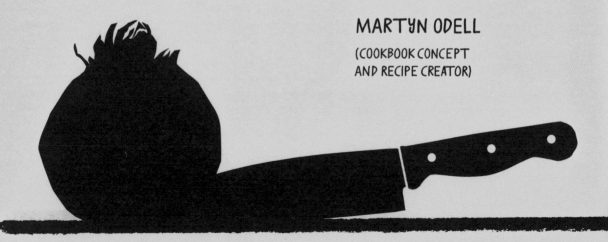

HELLO,

I've been creating recipes for Oddbox since 2018, showcasing how to make the most of rescued fruit and veg and sharing all my low-waste cooking tips along the way.

My speciality is in Provençal cuisine and the Mediterranean diet, and I have over 15 years' experience in the hospitality industry as a chef, recipe developer and cookery teacher. I'm also the founder of Call her Chef, a female-led cookery business.

CAMILLE AUBERT

(DESSERT MASTERMIND)

CRUNCH TIME.

Every year, about 2.5 billion tonnes of food goes to waste. And when food is wasted, so is all the energy and water used to grow and produce it. Most people don't realise that food waste is a climate issue, and one of the most urgent ones at that.

A lot of what we do at Oddbox focuses on tackling food waste on farms. We give growers a new market for crops that are at risk of going to waste - because they look 'too odd', because there are 'too many', because the weather's done something unexpected (surprise, surprise), because demand dips, because a harvest is ready earlier or later than expected . . . The list goes on.

But while a humongous quantity of food is lost in parts of the food system that most of us never see, the truth is that about 70% of food wasted beyond the farm gate is wasted at home. That's not to shift the responsibility away from a flawed food system to the average person doing their best - more to acknowledge that this is a problem we can all do something about.

In fact, food waste has been identified as one of the most fixable issues to help reverse the climate crisis - yes, you read that right. It's fixable. If we all make small changes to the way we shop, cook and eat, the knock-on effects could be massive - almost as massive as some of the peppers we saw last spring. (We said almost.)

WHY THE WASTE?

We've all been there. You get home from work. Have a quick glance in the fridge. Decide there's nothing you can make with a lonely-looking aubergine and you order a takeaway. The next day, you remember that aubergine and buy more food to make sure you have all the ingredients needed for the recipe you found online. One aubergine - tick. Ten new ingredients? Why not?

While making your new dish, you realise that one of your freshly-bought tomatoes has been squashed in the packet, so you throw it out. You already had a jar of oregano lurking at the back of the cupboard, so now you have two. The recipe only calls for half a courgette, so the other half is dropped into the salad drawer to await its - inevitable - fate. And all the odds and ends, outer leaves and nubbins are chucked out along the way.

A STAGGERING 2.5 BILLION TONNES OF FOOD GOES UNEATEN EVERY YEAR.*

ABOUT 70% OF FOOD WASTED BEYOND THE FARM GATE IS WASTED AT HOME.**

Although food waste is being talked about more and more, changing ingrained habits like these is harder than it seems. Which means we're buying more food than we need, we're spending money that could be spent on something else and we're cramming our cupboards and fridges with ingredients that will likely be forgotten about and thrown away.

Our community tells us that our boxes have helped to shift mindsets from 'what do I want?' to 'what do I have?'. Being faced with a surprise selection of fruit and veg each week encourages us to be resourceful and creative in our cooking, and to use what we have before buying anything else.

With a bit of planning, an open mind and a few adaptable recipes in our repertoire, we can all eat the food we have. We can look ahead, finding ways to preserve and store anything we won't have time to eat before it goes off. And we can learn how to make seeds, peel and leftovers go further. (Who knew you could make vinegar from fruit scraps?)

Best of all, we can start making these little changes right away. You don't need to buy anything new – all you need to cook creatively, waste less food and make a positive difference to the planet is in your kitchen and in this book.

It's crunch time.

*WWF, Driven to Waste Report, 2021
**WRAP, Food surplus and waste in the UK – key facts, updated Oct 2019

HOW THIS BOOK HELPS.

Most recipe books tell you what you need; this one helps you make the most of what you have. With a slight shift in thinking and a shake-up of our kitchens, we can all get more from our food, enjoying the process at the same time.

This book has three sections which will help you:

1. SET UP A WASTE-FREE KITCHEN.

Sort your kitchen cupboards, make your fridge and freezer work harder and store and preserve ingredients so they last longer.

2. SWAP, SWITCH & FREESTYLE.

Learn how to swap ingredients like a pro so you can create your own dish from scratch, based on what you already have.

3. COOK CREATIVELY.

Have a go at our delicious, flexible recipes – each one has a list of ingredients you can swap in and out so you can always use what's available.

Ready, steady, veg? Let's go.

PERFECT DOESN'T GROW ON TREES

1. SET UP A WASTE-FREE KITCHEN.

HOW TO GET ORGANISED AND
EAT MORE OF WHAT YOU'VE GOT.

MAKE SPACE TO COOK CREATIVELY.

One of the main reasons food is wasted at home is because it goes off before we have a chance to eat it. With just a few unexpected dinners out or a fussy eater in the family, salad leaves go limp, spuds start sprouting, carrots lose their crunch – and suddenly you've wasted a whole load of food without noticing.

For some people, planning meals for the week helps with this. But even if you're more likely to rustle something up than stick to a daily spreadsheet, some simple changes to the way your kitchen is organised can make a big difference in the fight against food waste.

Part of this is making sure your fridge and freezer are pulling their weight. Part of it is making sure you have the cupboard ingredients you need to whip up a variety of dishes. An even bigger part of it is making sure you can see what you have, so you can spot – and save – the ready-to-wilt before it flops.

Cooking should be an easy and fun activity, so you want to make sure you can chop, stir and sizzle without having to rummage around for the right pan or ingredient.

Martyn Odell recommends setting yourself up by creating an imaginary heat map of your kitchen. What do you use most? Can you find it quickly? What could you clear out to make it easier to reach when you're cooking and in full flow?

Keep all your favourite pans, bowls and colanders close to where you cook and pop your best wooden spoons, whisks and spatulas in a pot on the work surface. (Don't pretend you don't have a best wooden spoon – we all do.)

Then clear out anything you haven't used for yonks. Slow cooker gathering dust? Give it to a neighbour. Fifteen unopened boxes of lasagne sheets? List them on a food-sharing app. Herb jars still sporting a label from the 1970s? Compost anything ancient and recycle the containers. That's better.

FOOD-WASTE FIGHTING EQUIPMENT.

A few handy tools can make storing and preserving food that bit easier, not to mention giving you the freedom and flexibility to have a go at any recipe you fancy.

BLENDER

Make sauces, soups, smoothies, salsas (and other things that don't begin with s) – in a flash. You can even use a blender to chop ingredients if you're in a hurry.

AIRTIGHT JARS

Not only will a row of jars make your kitchen look like a TV chef's, they're also the best containers for making pickles and preserves.

LOCKABLE CONTAINERS

Leftovers, packed lunches, salad leaves – invest in lockable, leak-free boxes and you'll be able to keep all kinds of food fresh for longer. If they stack neatly, even better.

OVEN DISHES

This might sound obvious, but a few good ovenproof dishes mean you won't have to rehome leftover pies, lasagnes and bakes when you transfer them to the fridge.

REUSABLE FREEZER BAGS

Storing soups, stock or stews in the freezer? Good-quality, leakproof bags can be laid flat to give you extra space – and washed to use again and again.

PESTLE & MORTAR

Only have seeds when the recipe calls for powder? Bash, mash and squish things up yourself - it's good stress relief, too.

KITCHEN CADDY

Invest in a small bin for pips and peel on your kitchen counter (unless you're using them in a Cooking scrappy recipe - see pages 178-193). Then you can carry it straight to the compost.

And if you're looking for things for your birthday list...

DEHYDRATOR

Dehydration is one of the oldest methods of preserving food. These counter-top boxes sap the moisture out of everything from fruit and veg to herbs and meat (you can also use your oven - see page 28). Banana chips, anyone?

VAC-PAC MACHINE

Most food can be vacuum-sealed to keep it fresh for longer. Best of all, you can pack and label the right-sized portions for the freezer, so you only defrost as much as you need.

CHILLED TO ZERO- WASTE PERFECTION.

Your fridge and freezer are your best friends when it comes to tackling food waste. But – here's the snag – they won't be able to help if they're rammed with mystery jars, hidden tubs and unloved salad leaves. (Sorry to all the hummus-hoarders out there.) Start by taking everything out so you know what's what.

Then have a look at our tips for arranging things clearly and neatly. Your fridge should act as a mini shop, showcasing what you have and alerting you to anything that's about to pass its best. Then your freezer is there to store things so they can be eaten at a later date, locking in flavour at the same time.

FREEZER

SET THE TEMP TO -18C (-0.4F).

USE FREEZER BAGS SO YOU CAN LIE EVERYTHING FLAT.

AVOCADO CHUNKS 4/4

SQUASH PEELS 21/1

LEMON PEELS 16/1

SAVE PEELS, SCRAPS AND ENDS IN A BAG TO USE IN A STOCK.

WRITE THE NAME OF THE DISH AND DATE ON LEFTOVERS.

SQUASH CURRY 23/3

BOLOGNESE 23/3

FILL ICE CUBE TRAYS WITH LEFTOVER WINE AND STOCK FOR SAUCES, THEN TRANSFER TO A BAG WHEN FROZEN.

DID YOU KNOW YOU CAN FREEZE COCONUT MILK AND TOMATO PUREE?

KEEP FOOD FROZEN FOR UP TO THREE MONTHS. ONCE IT'S DEFROSTED, DON'T REFREEZE.

FREEZE BERRIES AND CHOPPED FRUIT ON TRAYS TO STOP THEM CLUMPING TOGETHER, THEN MOVE TO A BAG.

FRIDGE

SET YOUR FRIDGE TO UNDER 5C (41F).

ROTATE THINGS REGULARLY SO YOU CAN SPOT IF ANY LURKERS NEED EATING.

C&C SOUP 22/5

KEEP CONDIMENTS IN THE DOOR.

TOP AND MIDDLE SHELVES ARE FOR FOOD THAT'S READY TO EAT – LIKE LEFTOVERS AND DAIRY.

ANYTHING COOKED WILL STORE FOR UP TO 3 DAYS.

VEG CHILLI 23/5

UNLESS SOMETHING HAS A USE-BY DATE (WHICH SHOULD BE STUCK TO), USE YOUR INTUITION TO CHECK YOUR FOOD. IF IT LOOKS, SMELLS AND FEELS OKAY, IT'S PROBABLY FINE TO EAT.

BOTTOM SHELVES ARE FOR RAW MEAT AND FISH (STUFF THAT CAN DRIP AND SPREAD BACTERIA).

BACON / SAUSAGES 23/5

VEG DRAWERS ARE FOR... VEG!

NOT GOING TO EAT SOMETHING? FREEZE OR PRESERVE IT (SEE PAGES 27-31).

WASTE-FREE CUPBOARDS.

What have cupboards got to do with food waste, we hear you cry? Well, while it's much easier to waste fresh stuff, the way we store our dry food - and even what we buy - makes a big difference to how easily we can cook creatively and resourcefully.

With a good store of jars, tins, herbs and spices, you'll be able to whip up a delicious dish that makes use of your fresh ingredients - possibly even without a recipe. And if you get everything organised, there'll be less chance of buying more than you need.

MAKE THINGS VISIBLE

Like an artist choosing paint, it's nice to have your ingredients lined up somewhere you can see them clearly. If there's anything you want to use but keep forgetting about, move it up front.

STORE STUFF PROPERLY

If you buy loose ingredients, store everything in clear, airtight containers with labels. And make it easier to find things by grouping them in boxes or crates - or by labelling the shelves themselves.

LOOK BEFORE YOU SHOP

Simple one, this. But just checking what you have before you head to the shops or fill your basket will stop you wasting money and space. In restaurants, this kind of inventory happens at the end of every shift - Martyn recommends a Sunday night check.

KEEP REJIGGING

Before you unpack your shopping, reach to the back of cupboards and pull any hidden ingredients to the front. And put the new things further back, so you're always using up what you've got.

DO HERBS & SPICES GO OFF?

In a word: no. But after a while they lose their pizzazz, so only buy them in small amounts and check your cupboards before buying more.

MARTYN'S CUPBOARD CHECKLIST.

Having a good range of dry ingredients will make creative, no-waste cooking easier. Our list is based on the recipes in this book and what we use the most. But don't feel you have to buy things for the sake of it – stock up on whatever ingredients and flavours you like best.

SPICES & SEASONING

1. BLACK PEPPERCORNS
2. CARAWAY SEEDS
3. CARDAMOM
4. CAYENNE
5. CHILLI FLAKES
6. CINNAMON
7. CORIANDER SEEDS/GROUND
8. CUMIN SEEDS/GROUND
9. GARAM MASALA
10. MUSTARD SEEDS
11. NUTMEG
12. OREGANO
13. SEA SALT
14. SMOKED PAPRIKA
15. TURMERIC

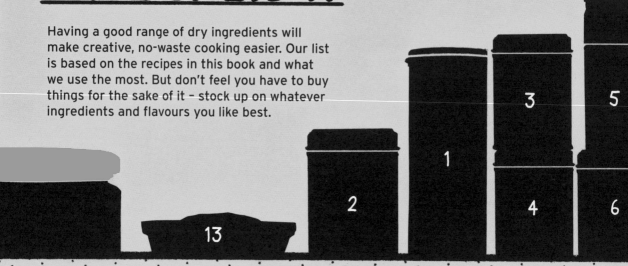

TINS

19. CHICKPEAS
20. COCONUT MILK
21. SWEETCORN
22. CHOPPED TOMATOES
23. BEANS (HARICOT, KIDNEY, PINTO ETC.)

JARS

24. MISO
25. PESTO
26. CURRY PASTE
27. MUSTARD

LIQUIDS

28. HOT SAUCES
29. SOY SAUCE
30. VINEGAR (RED, WHITE, BALSAMIC)

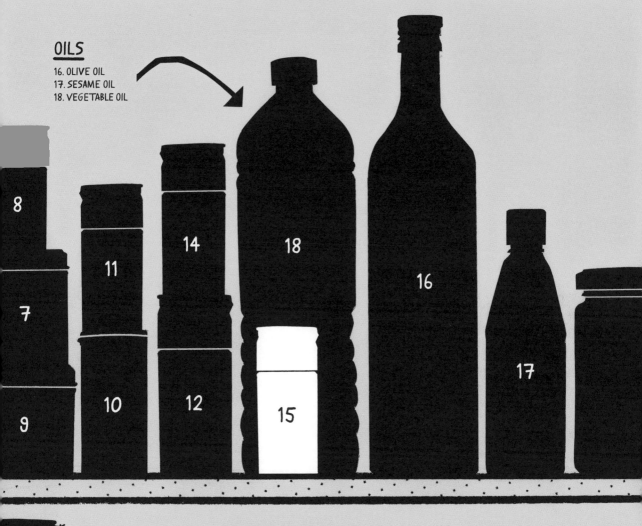

OILS

16. OLIVE OIL
17. SESAME OIL
18. VEGETABLE OIL

8
11
14
18
7
10
12
15
9
16
17

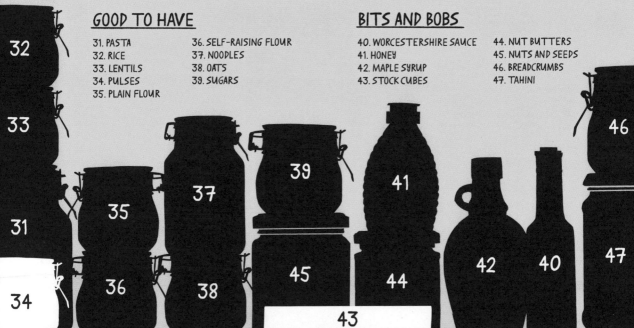

GOOD TO HAVE

31. PASTA
32. RICE
33. LENTILS
34. PULSES
35. PLAIN FLOUR
36. SELF-RAISING FLOUR
37. NOODLES
38. OATS
39. SUGARS

BITS AND BOBS

40. WORCESTERSHIRE SAUCE
41. HONEY
42. MAPLE SYRUP
43. STOCK CUBES
44. NUT BUTTERS
45. NUTS AND SEEDS
46. BREADCRUMBS
47. TAHINI

32
33
31
34
35
36
37
38
39
45
41
44
43
42
40
46
47

SIMPLE WAYS TO PRESERVE & PROLONG FOOD.

Don't wait until fresh food wilts – if you're not going to get through all your fruit and veg, either prep and store it using one of these cooking techniques, or turn it into a chutney or pickle. Ta daah!

BLANCHING & REFRESHING

Keep veg fresh for longer (and get ahead if you're prepping for a big meal). The cold water 'shocks' and halts the cooking process, so the veg keeps its colour, crunch and flavour until you're ready to reheat and eat.

WORKS FOR...
Asparagus, broccoli, Brussels sprouts, cauliflower, carrots, kale, spinach, cavolo nero, peas, cabbages, green beans, mangetout, herbs.

HOW?
Plunge your prepped veg into boiling water for 1–2 minutes (depending on the size), until very slightly softened. Then scoop them out and plunge straight into ice-cold water.

STORAGE
Keep in the fridge for 3–5 days, or in the freezer for three months.

MASHING & PURÉEING

Fruit and veg looking a bit bashed-up? Turn them into a mash or purée and no one will ever know. Savoury mashes can be used as a side dish or pie topping, while fruit purées can be used in desserts, to sweeten a curry or even to replace oil and butter in baking.

WORKS FOR...
Almost anything – try potatoes, apples, squash, cauliflower, beetroot, parsnips, carrots, celeriac.

HOW?
Boil, roast or steam your ingredients then mash them to a rough or smooth purée.

STORAGE
Keep in the fridge for 3–5 days, or in the freezer for three months.

BLITZING

Make life easier for future you by creating a frozen store of chopped ingredients to use in soups, stews, dips and salads (see page 79). Blitzed cauliflower is also great as a rice alternative.

WORKS FOR...
Onions, garlic, ginger, mushrooms, aubergines, cauliflower, broccoli, fresh herbs, nuts, seeds, oats. (Onion, garlic and ginger can be frozen raw - cook everything else first.)

HOW?
Use a blender, food processor or grater to chop everything into tiny pieces.

STORAGE
Store in the fridge for 3-5 days, or in the freezer for three months.

DEHYDRATING

By reducing the water content, dehydrating gives fruit and veg a much longer shelf life. You can make your own fruit leathers, chips and snacks. Or dehydrate savoury things such as mushrooms, celeriac and greens, ready to be rehydrated, used as a garnish or turned into a powder.

WORKS FOR...
Almost anything - try apples, pears, bananas, pineapple, berries, mangoes, tomatoes, mushrooms, celeriac, courgettes, onions, potatoes, kale, citrus peel.

HOW?
If you have a dehydrator, check the manual for timings. Or you can use the oven - set it to its lowest temperature (about 50-60°C), place your ingredients in a single layer on a baking tray and dry for 3-10 hours. You're aiming for a shrivelled, chewy texture with no stickiness.

STORAGE
Dehydrated fruit and veg keep for months in an airtight container.

CHUTNEY-MAKING

The word chutney is used for any fruit or veg preserved with vinegar, sugar and spices. Spend an hour making it, then add a tangy dollop to your meals for months to come.

WORKS FOR...
Almost anything – try apples, onions, pears, mangoes, raisins, tomatoes, plums, carrots, beetroot.

HOW?
Try to create a balance of textures – fruity ingredients will break down more to give a smoother consistency, while onions will keep their crunch. You can also mix up the flavours – try chilli, cardamom or mellow Indian spices.

STORAGE
Chutneys keep in a cupboard unopened for months. Once opened, keep the jar in the fridge and eat within a couple of weeks.

APPLE CHUTNEY

PREP TIME: 10 MINUTES
COOK TIME: 60-90 MINUTES
MAKES: 3-4 SMALL JARS

5 apples, cored and roughly chopped (500g)
3-4 onions, sliced (500g)
350ml any good-quality vinegar
350g sugar
1 cinnamon stick
4 cloves

SWAPS
No apples? Use pears, plums, apricots, tomatoes, bananas.

1. Put the apples and onions into a large saucepan.
2. Add the vinegar, sugar, cinnamon stick and cloves.
3. Bring the mixture to a boil, then allow to simmer on a low heat for 60-90 minutes, stirring occasionally until you have a thick and jammy consistency.
4. Allow to cool then pop into sterilised jars (see page 31).

CAN I EAT IT?

THESE ARE OUR TOP TIPS FOR FLOPPY, SQUISHY OR WRINKLY FRUIT AND VEG.

Potatoes with a few sprouts are still fine (just cut them off) but compost any that are mouldy, very green or soft.

Vegetables such as celery, spring onions, carrots and beetroot tend to lose moisture and look a bit floppy as they get older. If you want to eat them raw, pop them in a bowl of water to perk them up. Or just roast, blitz or purée.

Mash squidgy avocados into a guacamole, blend into a smoothie or freeze to use at a later date. (The brown bits might look unappetising, but they're still safe to eat.)

Wilted lettuce can be revived by standing it in cold water, but chuck it in the compost if it's brown or soggy.

If your fruit is bruised, cut out the brown parts and eat the rest. Or use it in a purée, pudding or cake. Compost anything that's really soft and mouldy.

PICKLING

Pickling is an art – choose your ingredients, experiment with different vinegars and play around with herbs and spices. When you're done, you'll have jars of delicious delicacies to chop up into salads or to use as side dishes.

WORKS FOR...
Almost any fruit or veg. Try carrots, parsnips, cucumbers, red onions, asparagus, strawberries, plums and apricots.

HOW?
A basic rule of thumb is to use equal parts water and vinegar. Beyond that, you can add extra flavours to personalise your pickles. Try dried spices such as turmeric or chilli powder, fresh herbs such as dill, garlic and ginger or whole spices including mustard seeds or coriander seeds.

STORAGE
Jars of pickle keep for months in the cupboard. Once you've opened a jar, store it in the fridge for up to seven days.

HOW TO?

STERILISE A JAR

You don't need special preserving jars to make chutneys and pickles, but you do need to make sure that whatever you use has an airtight lid. You also need to sterilise your jars before filling them, to kill any bacteria.

1. Wash the jars and lids in warm soapy water then rinse them.
2. Place them in a deep pan and cover with boiling water. Simmer for 10 minutes.
3. Take the jars out and let them cool before spooning in your mixture.
4. Put on the lids and flip the jars upside down to create a seal.

PICKLED CARROTS

PREP TIME: 10 MINUTES
REST TIME: 4 HOURS
MAKES: 2 JARS

4 carrots (230g)
100ml cider vinegar
100ml water
2 bay leaves
1 tsp mustard seeds
1 tsp black peppercorns
1 tsp sugar
pinch of salt

SWAPS
No carrots? Use sliced fennel, beetroot, onion, cucumber or cauliflower florets.

1. Chop the carrots - unpeeled - into sticks so they fit into your jars.
2. Pop the carrots in the jars, standing upright.
3. Put the vinegar, water, bay leaves, mustard seeds and peppercorns in a saucepan.
4. Bring the vinegar mixture to the boil.
5. Add the sugar and a good pinch of salt then turn off the heat.
6. Carefully pour the mixture over the carrots until they're fully covered.
7. Allow the jars to cool to room temperature.
8. Pop in the fridge and leave for at least 12 hours and up to two months.

2. SWAP, SWITCH & FREESTYLE.

BUILD YOUR SKILLS – AND
CONFIDENCE – IN THE KITCHEN.

SWAP, SWITCH & FREESTYLE.

Ever scanned a list of ingredients, realised you're missing an essential pak choi and decided not to bother? Or worse, headed out to the shops and bought a pack of four when you only need one? We think restrictive, rule-book-style recipes are not only causing us all stress; they're also leading to food waste.

The recipes in this book are designed to be flexible – if you don't have a particular vegetable, there's always something else you can use in its place. But before we start swapping things, it's helpful to know how and why some ingredients work in a similar way (and why you might want to reconsider swapping spinach for potatoes in your mash, for example).

Let's arm ourselves with the skills – and guts – to use what we have, instead of buying more. Let's stop worrying about getting it right and give ourselves the freedom to do things our way. And let's stop imagining that chefs don't improvise and adapt as they go.

After all, what's the worst that can happen? The slightly unexpected and different-every-time is what we do best. Odd is good – you heard it here first.

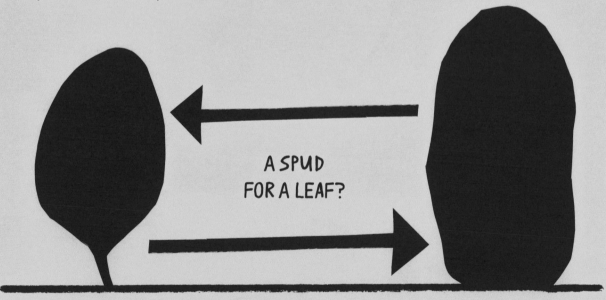

A SPUD
FOR A LEAF?

HAPPY FAMILIES.

Some ingredients cook in a similar enough way that you can swap them in and out without fuss. So while the flavour will obviously change slightly (yep, your gran's famous clementine cake won't taste *quite* the same if you swap in limes), the overall dish will still work perfectly.

MEET THE FRUIT & VEG YOU CAN EASILY SWAP. *

ALLIUMS & FRIENDS

LEEKS

ONIONS

SHALLOTS

FENNEL

SPRING ONIONS

FENNEL & SPRING ONIONS (DIFFERENT FAMILY, SAME COOKING VIBE).

ROCKET

BABY SPINACH

LETTUCE

WATERCRESS

SALAD LEAVES

BROCCOLI

CABBAGES

CRUCIFEROUS

BRUSSEL SPROUTS

CAULIFLOWER

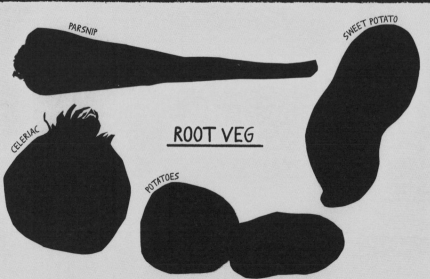

PARSNIP

SWEET POTATO

CELERIAC

ROOT VEG

POTATOES

APPLES

ORCHARD

PEARS

SQUASHY

PUMPKIN

SQUASHES

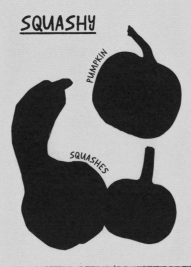

PEACHES

NECTARINES

STONE FRUIT

PLUMS

APRICOTS

CHARD

KALE

LEAFY
GREENS

SPINACH

CAVOLO NERO

BLACKBERRIES

RASPBERRIES

BERRIES

STRAWBERRIES

BLUEBERRIES

HONEYDEW MELONS

MELONS

WATERMELONS

LEMONS

CLEMENTINES

ORANGES

CITRUS

LIMES

GRAPEFRUITS

MANDARINS

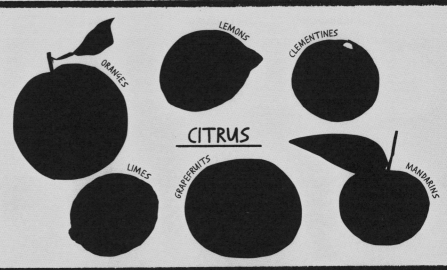

*THESE ARE MARTYN'S SUGGESTIONS. AS ALWAYS, THERE MAY BE SOME EXCEPTIONS TO THE RULES!

PLAYING SWAPSIES.

No peppers in the fridge? Married to a mushroom-hater? Little people refusing to eat peas? If you need - or want - to swap a fruit or veg in a recipe that doesn't have an obvious replacement, you'll have to think a bit more carefully about it.

First things first: ask yourself what role your ingredient plays - is it there to add flavour? Or is it fundamental to the structure of the dish? (As in, the dish would fall apart without it. Dramatic.) Once you've identified the role of your ingredient, you can then approach your swaps.

FLAVOUR SWAPS

Ingredients can add taste, flavour - and even texture - without being central to the way the dish works as a whole. For example, you can replace peppers in a stir-fry quite easily, swapping in anything you have in the fridge - mushrooms, courgettes, aubergine, etc. The cooking time will change slightly, along with the overall taste, but your dish will still work in much the same way.

EXAMPLES OF FLAVOUR SWAPS:

- Raspberries for pears in chocolate brownies
- Butternut squash for leeks in risotto
- Carrots for courgettes in a curry

STRUCTURAL SWAPS

Some ingredients are central to a recipe - they're so important that getting rid of them would mean that the dish doesn't hold together. Examples of this type of ingredient include potatoes in a croquette, apples in fritters or chickpeas in falafel. If you're going to swap a structural ingredient, you need to make sure the substitute is up to the job - for example, instead of potatoes, use another starchy root veg to keep your croquettes intact.

The fruit and veg families on page 36 will help you with this, as will the techniques on page pages 27–31 which show ingredients that can be cooked in a similar way.

EXAMPLES OF STRUCTURAL SWAPS:

- Parsnip for celeriac in a crusty pie topping
- Celeriac for potato in croquettes
- Pineapple chunks for sliced apples in fritters

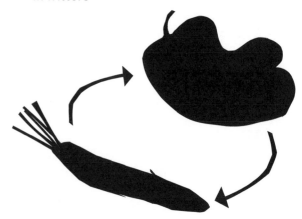

CAN I ROAST IT?

Not sure if you can swap chargrilled peppers for celeriac? These tables will help you swap fruit and veg like a pro - if you can't use the same technique for your substitute, cook it differently and tweak the recipe as you go.

The techniques and ingredients here are based on the recipes in this book and the most common fruit and veg in our boxes.

	BLANCH/BOIL	ROAST/BAKE	CHARGRILL	SAUTÉ	POACH	FRY	EAT RAW
APPLES		X	X	X	X	X	X
APRICOTS	X	X	X	X	X	X	X
BANANA		X	X	X	X	X	X
BLUEBERRIES	X	X			X		X
CLEMENTINES	X	X			X		X
GRAPES		X		X	X		X
KIWI		X			X		X
MANGO		X	X		X		X
MELON		X	X		X		X
NECTARINES	X	X			X		X
ORANGE	X	X		X	X		X
PASSION FRUIT							X
PEACHES		X	X		X	X	X
PEARS		X	X	X	X	X	X
PINEAPPLE		X	X	X	X	X	X
PLUMS		X	X		X	X	X
RASPBERRIES	X	X			X		X
RHUBARB	X	X	X	X	X	X	X
STRAWBERRIES	X	X			X		X

	BLANCH/BOIL	ROAST/BAKE	CHARGRILL	SAUTÉ	POACH	FRY	EAT RAW
AUBERGINE		X	X	X		X	
AVOCADO							X
BEETROOT	X	X		X	X	X	X
BROCCOLI	X	X	X	X	X	X	X
BRUSSELS SPROUTS	X	X	X	X	X	X	X
CABBAGE	X	X	X	X	X	X	X
CARROT	X	X		X	X	X	X
CAULIFLOWER	X	X	X	X	X	X	X
CELERIAC	X	X	X	X	X	X	X
CELERY		X		X		X	X
CHARD	X			X	X	X	X
COURGETTE		X	X	X		X	X
CUCUMBER							X
FENNEL		X	X	X	X	X	X
GREEN BEANS	X	X	X	X		X	X
KALE	X	X		X	X	X	X

	BLANCH/BOIL	ROAST/BAKE	CHARGRILL	SAUTÉ	POACH	FRY	EAT RAW
LEEKS	X	X	X	X	X	X	X
LETTUCE				X			X
MUSHROOMS		X	X	X	X	X	
ONIONS		X	X	X	X	X	X
PARSNIPS	X	X		X	X	X	
PEAS	X			X	X	X	X
PEPPER		X	X	X	X	X	X
POTATO	X	X		X	X	X	
RADISHES		X	X	X		X	X
SPINACH	X			X	X		X
SQUASH	X	X		X	X	X	
SWEDE	X	X		X	X	X	
SWEET POTATO	X	X		X	X	X	
SWEETCORN	X	X		X	X	X	X
TOMATO	X	X	X	X	X	X	X

WHAT'S THE BIG DILL?

Some recipes, such as pesto or salsa verde, need fresh
herbs to make them taste good. But others – where
you're just adding a sprinkling of flavour along the way
– are more flexible. Swap dried for fresh and fresh for
dried – use whatever you have at home.

HERB / SPICE	FRESH	DRY
BASIL	A HANDFUL	2 TSP
CHILLI	1 CHILLI	½ TSP CHILLI FLAKES
CINNAMON	1 STICK	½ TSP GROUND
DILL	A HANDFUL	1 TSP
OREGANO	A HANDFUL	1 TSP
ROSEMARY	A FEW SPRIGS	1 TSP
SAGE	A HANDFUL OF LEAVES	1 TSP
TARRAGON	A SMALL BUNCH	1 TSP
THYME	A FEW SPRIGS	1 TSP

THANKS!

Thanks to one of our
community members
for sharing a similar
chart – we owe you
big thyme.

DISH-IT-YOURSELF.

Shock, horror: recipe book tells you not to use recipes. Well, what's stopping you? Now that you have a rough idea of how different ingredients work, you can create your own dishes based on what you have and what you've made before. It's easier than you might think, and our dish builder is here to help set you off in the right direction.

1. PICK A HERO

Starting with your fridge, pick a fresh ingredient (or two) to be the main focus of your meal.

2. DECIDE ON A DISH

With your hero ingredient in mind, decide what you fancy. Some ideas: soup, salad, pasta sauce, stew, risotto, tray bake, curry, stir-fry, or sandwich.

3. ENLIST A CUPBOARD SIDEKICK

Every hero needs a sidekick. Pick a companion for your fresh ingredients from the cupboard - for example, rice, couscous, bread, pasta, noodles, beans, lentils, oats.

4. CHOOSE A TECHNIQUE

Decide which cooking method will work best for the dish you have in mind. For example, frying, roasting, pickling, mashing, chargrilling, sautéing, poaching. If you have a few fresh ingredients, don't be afraid to cook each one individually to bring out their flavours - roasting peppers in the oven while you sauté an onion, for example.

5. BALANCE THE FLAVOURS

Cooking is all about balance - sweetness to balance spice, salt to balance bitterness, sourness to enhance sweetness. With the aim of creating balance (i.e. not making everything taste the same), decide where you can add different flavours in your dish. Can you use a herb or spice? Would a dressing or sauce go with it? Could you pickle something to add tang? And if you add spice, what can you use to soften it?

6. ADD TEXTURE

Giving a dish texture helps to make it more interesting and enjoyable - think of silky soup with croutons on top. See if there's a way to add crunch to glossy textures, or smoothness to chunky ones.

7. TASTE TO SEE IF IT NEEDS A FINAL FLOURISH

A squeeze of lemon to balance salty tacos? A dollop of yoghurt to calm a fiery curry? Crispy onions, nuts or seeds to add a twist to a salad? Taste your dish, and add a final flourish if it needs one.

Then give your dish a name, serve and enjoy. Told you it was possible.

3. COOK CREATIVELY.

SWAP-ABLE, LOW-WASTE RECIPES FOR EVERY MEAL OF THE DAY.

COOK CREATIVELY.

No courgettes? No problem. We set Martyn Odell and Camille Aubert the challenge of creating a collection of delicious, swap-able recipes that would tackle food waste simply by encouraging us to use up what we've got. Along with new dishes, you'll also find some of our favourite Oddbox recipes from over the years, too.

The idea isn't that you should try to recreate the photos (although we know they look delicious). It's to start with our base recipe, then add whatever fruit and veg you have, using the suggested swaps as inspiration. And if it comes out a totally different colour because you swapped carrots for peppers? Even better.

If you have any leftovers, look for the info at the bottom of each page on how to store or transform what you've got (arancini balls from leftover risotto - magic). And if you're left with pips, skins and scraps, be sure to flick to page 178 for our Cooking scrappy recipes.

Spoons, spatulas and squashes at the ready - let's cook like there is a tomorrow.

A NOTE ON SWAPS

Each recipe has a list of fruit and veg swaps, separated into Flavour swaps and Structural swaps (flick to page 38 for more on this). Some swaps work in exactly the same way as the original ingredients, but with others, you might have to make a few adjustments to the method. Oh, and if there's an ingredient you don't have or like which isn't central to the dish? Feel free to leave it out altogether. No one will ever know.

VEGGIE OR VEGAN?

Some of the recipes are vegan to start with - look out for the little VE at the top of the page. And lots of the others can be adapted with vegan substitutes. (If you find an alternative for eggs in frittata, though, we're all ears!)

LOOK OUT FOR ME

MESSAGE
FROM MARTYN.

'EVERY RECIPE IN THIS BOOK HAS BEEN DESIGNED TO BE AS FLEXIBLE AS POSSIBLE, SO GET CREATIVE AND SWAP INGREDIENTS BASED ON WHAT YOU HAVE. FEEL FREE TO DIAL UP THE FLAVOURS TO SUIT YOUR OWN TASTES, AND DON'T WORRY IF YOUR VERSION COMES OUT LOOKING DIFFERENT TO THE PICTURE – THAT'S ALL PART OF THE FUN. FINALLY, IF YOU ADD SOMETHING THAT WORKS WELL (OR SOMETHING THAT DOESN'T!), BE SURE TO MAKE A NOTE OF IT. THAT WAY, YOU'LL END UP WITH YOUR OWN COLLECTION OF RECIPES THAT YOU CAN COME BACK TO TIME AND TIME AGAIN.'

BREAKFAST.

DISHES WORTH LEAVING YOUR DUVET FOR.

56

61

69

64

VEGGIE HASH WITH EGGS.

PREP TIME: 5 MINUTES
COOK TIME: 15 MINUTES
SERVES: 2

1 large potato (450g)
100g kale, chopped
6 chestnut mushrooms (150g)
40ml olive oil, plus extra for
  frying the eggs
1 tsp smoked paprika
2 garlic cloves, thinly sliced
2 eggs
handful of fresh herbs
salt and pepper

SWAPS
FLAVOUR
• No mushrooms? Use sliced peppers, diced courgettes, diced aubergine.
• No kale? Use spinach, cavolo nero, spring greens.
STRUCTURE
• No potatoes? Use sweet potatoes, celeriac, parsnips, new potatoes, butternut squash.

LEFTOVERS?
THE COOKED MIXTURE WILL KEEP IN THE FRIDGE FOR THREE DAYS OR IN THE FREEZER FOR THREE MONTHS. REHEAT IN A PAN OR THE OVEN.

Crispy potatoes for breakfast? Yes please. The beauty of this dish is that you can add almost any veg you like – check the fridge, chop it all up and go wild. (And if you went wild the night before, you won't find a better meal for the morning after.) You can even parboil the potatoes in advance, or blanch and refresh the kale (see page 27) if you have guests and want to get ahead.

1. Dice your potato into small chunks, skin-on, and drop into boiling water. Cook for 5 minutes, or until the edges are turning soft. Scoop them out and set aside. Blanch the kale for 1-2 minutes, then drain.
2. Slice the mushrooms in half. Place a frying pan over medium heat and add half the oil, then fry for 4-5 minutes, turning once or twice to add some colour. Season with salt and pepper. (If you move mushrooms about too much they leach liquid and lose their flavour.)
3. Meanwhile, in a separate large frying pan, heat the remaining oil, add the potatoes and fry for 10-15 minutes. When they're cooked through and crispy, add the paprika, garlic and kale.
4. Keep cooking for 1-2 minutes and season with salt and pepper.
5. Add the mushrooms to the potato and kale mixture.
6. Meanwhile, pour a little oil into the empty mushroom frying pan and heat over a low to medium heat. Crack the eggs, and gently fry them for 2-3 minutes or until the whites have just set.
7. Mix any fresh herbs you have (parsley, dill or coriander, for example) through the potato hash, lay your eggs on top and get stuck in.

KEEP YOUR GARLIC SKINS TO MAKE A STOCK – SEE PAGE 185.

VERY-VEGGIE SHAKSHUKA.

PREP TIME: 10 MINUTES
COOK TIME: 45 MINUTES
SERVES: 2

8 tomatoes (600g)
1 red pepper (220g)
4 tbsp olive oil
2 tsp cumin seeds
1 onion, thinly sliced
1 courgette (150g), grated
2 garlic cloves, sliced
1 tsp smoked paprika
4 eggs
handful of fresh coriander,
  roughly chopped
salt and pepper

SWAPS

FLAVOUR
• No onions? Use sliced fennel, leeks or shallots, or chopped celery.
• No courgettes? Use diced aubergine, grated cauliflower, grated carrots.

STRUCTURE
• No fresh tomatoes? Use tinned tomatoes.

LEFTOVERS?
THE COOKED MIXTURE WILL KEEP IN THE FRIDGE FOR UP TO FIVE DAYS, OR IN THE FREEZER (ONCE COOL) FOR THREE MONTHS.

This is Martyn's twist on a classic Ottolenghi recipe - packed with flavour, but slightly simplified and with a few more vegetables thrown in for good measure. If you're cooking it at the weekend, take your time with the sauce and make a big batch to keep for later. Or for a quicker brekkie, sauté the peppers and use tinned tomatoes instead of roasted ones - it will still taste phenomenal.

1. Preheat the oven to 220°C/200°C fan/425°F/gas 7.
2. Slice the tomatoes and pepper in half, removing the seeds and stalk from the pepper. Drizzle with 2 tablespoons of olive oil and season with salt and pepper. Whack them on a roasting tray, cut side down.
3. Roast for 15-20 minutes, until the edges have started to catch and burn. This roasting will deepen the flavours of the sauce.
4. Meanwhile, put the remaining 2 tablespoons of olive oil in a large saucepan over a medium heat, add the cumin seeds and fry for 2 minutes.
5. Add the onion slices and grated courgette and fry for 5-10 minutes until super soft and caramelised.
6. Turn down the heat and add the sliced garlic and paprika. Fry for another 1-2 minutes.
7. Take the roasted tomatoes and peppers out of the oven and carefully slide them into the pan.
8. Bring to a simmer and roughly mash the ingredients with the back of a fork. Then add a little water and keep simmering for 10-15 minutes until the sauce looks like a thick soup.
9. Taste and season with salt and pepper, then make four wells with the back of a spoon and crack an egg into each.
10. Pop on a lid and let the eggs steam for 8-10 minutes on a very low heat.
11. When the eggs are ready, scatter over fresh coriander and serve with crusty bread.

ROASTED-VEG BAGELS.

PREP TIME: 10 MINUTES
COOK TIME: 15 MINUTES
SERVES: 2

1 courgette (160g)
1 aubergine (250g)
3 tbsp olive oil
2 garlic cloves, grated
zest of 1 lemon
200g halloumi or vegan
  alternative
2 bagels, or toast, wraps or
  breakfast muffins
handful of rocket
salt and pepper

SWAPS
FLAVOUR
• No courgette or aubergine?
Use red onion wedges,
mushrooms, sliced peppers,
broccoli florets, asparagus.
• No rocket? Use lettuce, salad
leaves, baby spinach.

Love bagels? These are crammed with roasted veg - perfect for prepping in advance and leaving in the fridge. Swap for toast, wraps or breakfast muffins if you prefer. And if you've got leftovers of the Chargrilled veg with whipped feta recipe on page 140, even better.

1. Get the chargrill pan screaming hot.
2. Cut the courgette and aubergine lengthways into ½cm thick pieces and place them in a bowl.
3. Drizzle with half the oil, season with salt and pepper and lay on the chargrill pan for about 2-3 minutes on each side.
4. While the courgette and aubergine pieces are grilling, place the grated garlic, lemon zest and a little oil in a bowl. Add the grilled veg and set aside.
5. Slice the halloumi into 2cm thick pieces and cook in the same pan over medium heat for 2-3 minutes on both sides, until crispy on the outside and soft in the middle.
6. Cut the bagels in half and toast. Then fill with the chargrilled veg, halloumi and rocket and get stuck in.

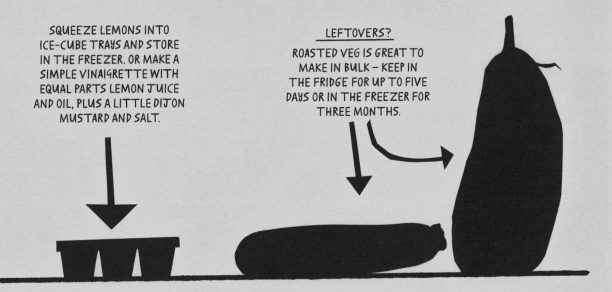

SQUEEZE LEMONS INTO
ICE-CUBE TRAYS AND STORE
IN THE FREEZER. OR MAKE A
SIMPLE VINAIGRETTE WITH
EQUAL PARTS LEMON JUICE
AND OIL, PLUS A LITTLE DIJON
MUSTARD AND SALT.

LEFTOVERS?
ROASTED VEG IS GREAT TO
MAKE IN BULK – KEEP IN
THE FRIDGE FOR UP TO FIVE
DAYS OR IN THE FREEZER FOR
THREE MONTHS.

BAKED CHOCCY-ORANGE PORRIDGE.

PREP TIME: 10 MINUTES
COOK TIME: 15 MINUTES
SERVES: 2

1 orange (100g)
70g rolled oats
1 tbsp chia seeds
1 tbsp cocoa powder
1 tbsp nut butter
1 tsp vanilla paste
½ tsp ground cinnamon
1 tbsp maple syrup
150ml any milk

SWAPS

FLAVOUR
• No orange? Use clementines, tangerines, bananas, kiwis, strawberries, blueberries.

We like to prep these oats the night before, then pop them in the oven while the kettle boils for breakfast. Best of all, the only bit of the orange you don't use is the pith - a waste-free start to the day. If you swap another fruit in, chop it up into even-sized chunks instead of squeezing and zesting. (You'll struggle with a banana - trust us.)

1. Preheat the oven to 200°C/180°C fan/400°F/gas 6.
2. Zest the whole orange and place it in a mixing bowl. Then cut off the top and bottom, and carefully cut off the peel, following the shape of the orange.
3. Holding the orange over a bowl, cut out the segments with a knife, catching the juice in the bowl. Put your segments to one side and squeeze all the juice from what's left.
4. In a mixing bowl, combine the orange zest, oats, chia seeds, cocoa powder, nut butter, vanilla paste, cinnamon, maple syrup, milk and a splash of the orange juice (drink the rest).
5. Mix well with a wooden spoon and leave the mixture to thicken for 5 minutes, then stir in half the orange segments.
6. Pour the mixture into a small baking dish and bake for 10-15 minutes until bubbling.
7. When ready, top with the remaining orange segments and dish up with a dollop of yoghurt, if you have any.

LEFTOVERS?
KEEP ANY EXTRA OATS IN THE FRIDGE FOR UP TO THREE DAYS.

THIS RECIPE USES ALMOST EVERY BIT OF THE ORANGE, BUT YOU CAN ALSO USE THE LEFTOVER PEEL IN MULLED WINE.

ROASTED-PEPPER FRITTATA.

PREP TIME: 15 MINUTES
COOK TIME: 40 MINUTES
SERVES: 2

2 peppers, any colour (440g)
5 eggs
100g spinach, plus extra to
    serve (optional)
handful of fresh herbs: parsley,
    basil, oregano (optional)
20ml olive oil
salad leaves, to serve
salt and pepper

SWAPS
FLAVOUR
• No peppers? Use diced
aubergine or courgettes, sliced
onion, fennel or mushrooms, or
mini cauliflower florets.
• No spinach? Use sliced
asparagus, defrosted garden
peas, rocket.

Why have an omelette when you can have a frittata? The difference is all in the making - unlike omelettes, frittatas are finished off in the oven, giving a lovely baked-egg taste. This recipe uses roasted peppers, which you can cook in bulk and keep for future recipes. Or, if you want a speedier breakfast, you can also sauté them in strips on the hob.

1. Preheat the oven to 180°C/160°C fan/350°F/gas 4, and turn the grill to high.
2. Put the peppers under the grill - whole - until they start popping and blistering. Turn and repeat until the skin is black and blistered all over. (A BBQ will give the same result.)
3. When they're blackened, pop the peppers in a bowl and cover with a tea towel until they are cool enough to handle.
4. Gently pick away the burnt skin under running cold water. Slice the peppers into strips.
5. Crack the eggs into a bowl and whisk, then add the pepper slices along with the spinach and any fresh herbs. Season with salt and pepper.
6. Heat a non-stick pan on a medium heat, drizzle in the olive oil then pour in the frittata mix. Gently fry for 2-3 minutes to set the eggs.
7. Slide the pan into the oven for 8-10 minutes or until the egg is cooked all the way through.
8. Serve with a good handful of salad leaves or more spinach leaves.

LEFTOVERS?
THE COOKED MIXTURE WILL KEEP
IN THE FRIDGE FOR UP TO FIVE
DAYS, OR IN THE FREEZER FOR
THREE MONTHS (ONCE COOL).
REHEAT IN THE OVEN UNTIL IT'S
PIPING HOT IN THE MIDDLE.

PAT THE PEPPER
SEEDS DRY AND
TOAST IN A PAN
FOR A GARNISH.

BAKED-FRUIT CRUNCH.

PREP TIME: 10 MINUTES
COOK TIME: 20 MINUTES
SERVES: 2

2 rhubarb stalks (200g)
2 tsp sugar
30ml vegetable oil
100g oats
pinch of ground cinnamon
30ml maple syrup
40g any nuts or seeds
150g any yoghurt

SWAPS
FLAVOUR
• No rhubarb? Use wedges of apple, pear, peach or plum, or pieces of pineapple or banana.

Soft, warm fruit. Crunchy nuts and oats. And a big dollop of creamy yoghurt. We've written this sunny recipe for rhubarb, but if it's out of season or you have something different in the fruit bowl, it's also delicious with any other bake-able fruit. Make a big batch and brighten up breakfast for days.

1. Preheat the oven to 200°C/180°C fan/400°F/gas 6.
2. Cut the rhubarb into finger-length pieces and spread them in a single layer on a baking tray lined with baking paper.
3. Sprinkle the sugar on top so it just covers the rhubarb and bake for 10–15 minutes, or until a knife slides in easily.
4. While the rhubarb is cooking put the oil, oats and cinnamon in a frying pan and cook over medium heat, stirring constantly. The oats will crisp up after about 5 minutes – turn off the heat when they're nice and golden.
5. Add the maple syrup, nuts such as chopped almonds, or seeds and a pinch of salt. Give everything a good mix.
6. Scoop a generous dollop of yoghurt into a bowl, make a well and pop in the rhubarb. Scatter over the toasted oats and enjoy.

LEFTOVERS?
KEEP THE RHUBARB IN AN AIRTIGHT CONTAINER IN THE FRIDGE FOR UP TO FIVE DAYS OR IN THE FREEZER FOR UP TO THREE MONTHS. STORE THE TOASTED OATS IN AN AIRTIGHT CONTAINER FOR UP TO 14 DAYS.

KIWI PANCAKES.

PREP TIME: 10 MINUTES
COOK TIME: 10 MINUTES
SERVES: 2 (6 PANCAKES)

3 kiwis
150g self-raising flour
20g porridge oats
1 tsp chia seeds
1 tsp baking powder
200ml any milk
15ml olive oil
1 tbsp maple syrup
coconut oil, for frying

FOR THE TOPPING
maple syrup
2 tbsp flaked almonds

SWAPS
FLAVOUR
• No kiwis? Use pineapple, strawberries, mangoes, grated apple, whole blueberries, mashed banana.

VE

It's time to up your pancake game. This nutritious recipe uses chia seeds, oats and chopped kiwi (why not?), but you can swap in pretty much any fruit you like and use dairy or plant-based milk. You can even cook the pancakes in advance and freeze them – magical.

1. Slice off the kiwi ends and compost. Then chop them, skin on, into small cubes.
2. In a bowl, mix the flour, oats, chia seeds and baking powder. Make a dip in the middle and pour in the milk, oil and maple syrup.
3. Whisk everything together to form a smooth batter. Then stir through the chopped kiwis, keeping a tablespoon for the topping.
4. Warm a teaspoon of coconut oil in a pan over medium heat. Spoon 3 big tablespoons of pancake batter into the pan.
5. Cook on one side for 1–2 minutes. When bubbles start to appear, flip the pancake and cook on the other side for 1–2 minutes, until puffed and golden.
6. Repeat until the pancake batter is finished.
7. Stack the pancakes and serve with the rest of the chopped kiwi, a drizzle of maple syrup and flaked almonds. Delish.

LEFTOVERS?

KEEP YOUR PANCAKES IN AN AIRTIGHT CONTAINER IN THE FRIDGE FOR UP TO THREE DAYS. OR STACK THEM WITH A PIECE OF BAKING PAPER BETWEEN EACH ONE AND FREEZE FOR UP TO A MONTH.

MIDNIGHT OATS.

PREP TIME: 5 MINUTES
REST TIME: OVERNIGHT
SERVES: 2

2 bananas, peeled (150g)
100g blueberries
250ml any milk
2 tbsp peanut butter
1 tsp ground cinnamon
100g oats
20g flaxseed
pinch of salt (optional)

FOR THE TOPPING
40g any nuts or seeds
handful of blueberries

SWAPS
FLAVOUR
• No blueberries? Use
strawberries, raspberries,
peaches, grated apple.
STRUCTURE
• No bananas? Use mangoes.

Just as we deliver our fruit and veg boxes overnight to keep emissions low, you can save your own energy by getting breakfast ready before bed. The secret to this recipe is soaking the oats in a smoothie good enough to drink. As soon as you spot berries looking a bit mushy, pop them into a bag in the freezer so you'll always have a ready supply for the blender.

1. Blend the bananas, blueberries, milk, peanut butter and cinnamon in a food processor.
2. Place the oats and flaxseed in a bowl, then pour the smoothie mixture on top.
3. Mix everything together well. Have a little taste, and if needed, add a pinch of salt.
4. Scoop the mixture into jars and pop in the fridge overnight or for a couple of hours.
5. At breakfast, top with nuts, seeds and more blueberries.

LEFTOVERS?
KEEP YOUR OATS IN
THE FRIDGE FOR UP TO
THREE DAYS.

SPICY BEAN BREAKFAST STEW.

PREP TIME: 10 MINUTES
COOK TIME: 35 MINUTES
SERVES: 2

1 white onion, diced (150g)
2 carrots, grated (100g)
30ml olive oil
2 garlic cloves, sliced
1 red chilli, finely diced
1 piece fresh ginger (15g),
   grated
1 tin tomatoes (400g)
1 cinnamon stick
2 tsp ras el hanout
1 tin any beans, drained (225g)
salt and pepper

FOR SERVING
handful of fresh coriander,
   roughly chopped
1 avocado, sliced
100g feta cheese, or vegan
   alternative, crumbled

SWAPS
FLAVOUR
· No white onions? Use sliced
leeks, fennel, red onions, celery.
· No carrots? Use cauliflower,
courgettes, beetroot, sliced
mushrooms, diced aubergine.

You know how some dishes taste better the longer you leave them in the fridge? This is one of them. Use any beans you like. Cook everything nice and slowly. Then enjoy rich, scrummy leftovers that just get better and better as the flavours soak in.

1. Gently sweat the diced onion and grated carrots in olive oil in a large saucepan for 5-8 minutes, until soft and caramelised.
2. Turn down the heat and add the sliced garlic and chilli (with seeds, if you like it spicy). Add the grated ginger and fry for 2-3 minutes.
3. Pour in the tinned tomatoes, half a tin of water and the spices. Bring to the boil.
4. Add the drained beans to the sauce. Then reduce to a simmer and let it bubble away for 15-20 minutes until nice and thick.
5. Season with salt and pepper and add extra spice if you want it, then stir through some of the chopped coriander and simmer for a few minutes.
6. Scoop the bean stew into bowls and top with more chopped coriander, sliced avocado and crumbled feta. Yum.

LEFTOVERS?
KEEP YOUR STEW IN
THE FRIDGE FOR UP
TO THREE DAYS, OR IN
THE FREEZER FOR UP
TO A MONTH.

SMOOSHED-FRUIT MUFFINS.

PREP TIME: 10 MINUTES
COOK TIME: 30 MINUTES
SERVES: MAKES 12 MUFFINS

100g blueberries
20g caster sugar
100g any butter, melted
2 large eggs
100ml any yoghurt
125g soft brown sugar
250g self-raising flour
50g oats
1 tsp baking powder
1 tsp bicarbonate of soda
1 tsp ground cinnamon

SWAPS
FLAVOUR
• No blueberries? Use raspberries, blackberries, chopped strawberries, mangoes, apples, pears.

LEFTOVERS?
KEEP YOUR MUFFINS IN AN AIRTIGHT CONTAINER IN THE FRIDGE FOR UP TO FIVE DAYS, OR IN THE FREEZER FOR UP TO THREE MONTHS.

Blueberries. Raspberries. Mango. All of the above. The joy of these fruity muffins is that you can use any fruit you fancy – just make sure you cook everything down until you have a nice, soft compote. If you're baking for a party or brunch, the muffin batter can be kept in the fridge for up to five days, then mixed with the compote just before they go in the oven.

1. Preheat the oven to 180°C/160°C fan/350°F/gas 4.
2. Pop the blueberries in a saucepan with the sugar and a splash of water, and gently heat until they're soft and sticky. Keep the compote chunky if you prefer or cook it until super smooth.
3. Meanwhile, whisk together the melted, cooled butter with the eggs and yoghurt to make a wet mixture.
4. In another bowl, mix the rest of the dry ingredients. Then pour them into the wet ingredients and mix well to make a batter.
5. Gently stir the blueberry compote through the muffin batter to create a ripple effect.
6. Dollop into cupcake cases or a greased muffin tin (fill halfway to the top). Sprinkle a few oats on top.
7. Bake for 15-20 minutes until cooked through. Then enjoy.

DON'T WORRY IF THE MIXTURE GOES A FUNNY PURPLE COLOUR WHEN YOU STIR THROUGH THE COMPOTE – IT'S JUST THE BLUEBERRIES DOING THEIR THING.

LIGHT BITES & LUNCHES.

SALADS, SOUPS AND MORE.

76

82

78

88

LOADED FLATBREADS WITH HUMMUS.

PREP TIME: 15 MINUTES
COOK TIME: 40 MINUTES
SERVES: 2

FOR THE BEETROOT
2 big beetroot (250g)
olive oil, for drizzling
1 tsp ground cumin
30ml balsamic vinegar
salt

FOR THE FLATBREADS
120g plain flour
80g any yoghurt
1 tsp baking powder

TOPPINGS
handful of fresh coriander
onion pickles
80g hummus (page 138)

SWAPS
FLAVOUR
• **No beetroot?** Use aubergine, cauliflower, carrots, peppers, sweet potatoes.

LEFTOVERS?
THE FLATBREADS CAN BE MADE IN A BIG BATCH, ROLLED OUT, COOKED AND THEN FROZEN BETWEEN BAKING PAPER FOR THREE MONTHS. THEY'RE VERY QUICK TO DEFROST — JUST POP THEM IN THE OVEN FOR 1-2 MINUTES.

Roll them up. Load them high. Or just rip, dunk and scoop. Flatbreads are super-easy to make and can give any dish a little bit more oomph. This recipe uses the hummus from page 138, and you can also add pickled onions - see page 30.

1. Preheat the oven to 180°C/160°C fan/350°F/gas 4.
2. Give the beetroot a quick wash and cut into chunky, even wedges - about eight per beetroot. Drizzle with a little oil, ground cumin and salt. Then pop on a tray and slide into the oven for 20-30 minutes or until nicely roasted.
3. While the beetroot is roasting, make the flatbreads. Mix the flour, yoghurt and baking powder in a bowl. The mixture will feel a little dry but when you start kneading it will come together.
4. Tip the mix onto a clean work surface and knead with the heel of your hand for 2-3 minutes until you have a silky-smooth dough. Dust with a bit of flour if it's too sticky.
5. Cut the dough in two, then roll out the pieces with a rolling pin until they're the thickness of a pound coin.
6. Heat a pan over a medium heat, lay in the flatbread and cook for 1 minute on each side. Brush with a little oil after turning and cook until golden brown. Tip: lay the flatbreads inside a clean tea towel to keep them warm and soft enough to roll.
7. When the beetroot is almost ready, drizzle with balsamic vinegar and crank the oven up to 240°C/220°C fan/475°F/gas 9 for 5 minutes so it goes nice and sticky.
8. Decide whether you want to keep your flatbreads open or roll them up - either way, load them with the roasted beetroot, hummus, pickles and fresh coriander and tuck in.

YOU CAN EVEN USE YOUR FLATBREADS AS AN EASY PIZZA BASE!

BLITZED SALAD.

PREP TIME: 20 MINUTES
COOK TIME: 10 MINUTES
SERVES: 2

1 tsp cumin seeds
1 tsp caraway seeds
1 tsp fennel seeds
50ml olive oil
2 garlic cloves, thinly sliced
2 carrots (150g)
2 spring onions (40g)
200g Brussels sprouts
50g raisins
handful of fresh coriander
50g sunflower or pumpkin
  seeds
juice of 1 lemon

SWAPS

FLAVOUR
• **No carrots? Use cauliflower,
peppers, beetroot.**
• **No Brussels sprouts? Use
kale, spinach, cabbage.**
• **No spring onions? Use red
onions, white onions, fennel.**

VE

Blender to the rescue – this super-speedy dish is perfect if you
need to whip up a salad in a flash. Chuck in your veg. Toss in
your aromatic dressing and seeds. Give it a whizz, and lunch
is served.

1. Start by toasting your spices over a low heat in a frying pan
for 2–3 minutes. Give them a shake so they cook evenly.
2. Pour in the olive oil and add the garlic, then very gently fry
for 2–3 minutes or until the garlic is just softening. Set the oil
mixture to one side.
3. Pop the carrots, spring onions, sprouts, raisins and
coriander into the blender and blitz to a rough mix. (Don't
overdo it – you want the veg to have some crunch.)
4. Toast the seeds in a dry pan for about 1–2 minutes.
5. Scoop the blitzed veg into a bowl and pour over the spiced
oil mix and a sprinkling of toasted seeds. Dish up and squeeze
the lemon juice on top.

LEFTOVERS?
THE SALAD WILL KEEP IN A
CONTAINER IN THE FRIDGE
FOR UP TO THREE DAYS.

WHOLE CAULI & TAHINI SALAD.

PREP TIME: 20 MINUTES
COOK TIME: 20 MINUTES
SERVES: 2

1 cauliflower (400g)
olive oil, for drizzling
100ml white wine vinegar
20g sugar
salt and pepper

FOR THE DRESSING
50g tahini
1 garlic clove
small handful of fresh coriander
small handful of fresh parsley
juice of 1 lemon
40ml water

FOR THE SALAD
1 red onion (150g)
2 big handfuls of rocket
1 tin any beans, drained (225g)
20ml olive oil
1 tsp harissa paste (optional)
½ tsp curry powder (optional)

SWAPS
FLAVOUR
• No cauliflower for roasting?
Use broccoli, carrots, onions,
beetroot, fennel.
• No cauliflower for pickling?
Use carrots, beetroot, fennel,
onions.
• No onion? Use spring onions.
• No rocket? Use salad leaves,
watercress, radicchio.

Florets, leaves, stem and all. Cauliflowers are one of the most versatile vegetables – roast some bits, pickle others and create a super-delicious dish without wasting a scrap. Beans also play a starring role here, bringing both texture and flavour to the party – try using cannellini, pinto, chickpeas or kidney beans.

1. Preheat the oven to 200°C/180°C fan/400°F/gas 6.
2. Break your cauliflower into florets, stem and leaves. Drizzle the florets with olive oil and season with salt and pepper, then place on a baking tray and roast in the oven for 15-20 minutes until slightly crispy.
3. Meanwhile, shred the cauliflower leaves into very fine strips and cut the stalk into super-thin matchsticks. Set to one side. (About 120g is enough, so save any extra for another time.)
4. In a saucepan, bring the vinegar to a simmer then add the sugar and stir until it's all dissolved. Turn off the heat and pop in the cauliflower leaves and stems, then leave for 5-10 minutes.
5. To make the dressing, blitz the tahini, garlic, coriander, parsley and half the lemon juice in the blender. Add water little by little until you have a smooth pourable sauce, and season to taste.
6. For the salad, peel and slice the onion thinly and place in a bowl of cold water for 2-3 minutes to take out the punch.
7. Construct the salad by first dressing the rocket and some of the beans with the rest of the lemon juice, oil and salt (harissa paste optional here). Then randomly layer the cauliflower florets, pickled leaves and stalk, red onion, more beans and tahini dressing.
8. Finish with a good twist of salt and a dusting of curry powder (optional), and serve with a flatbread.

LEFTOVERS?

THE CAULIFLOWER WILL KEEP FOR THREE DAYS IN THE FRIDGE; THE TAHINI DRESSING WILL BE GOOD FOR UP TO FIVE DAYS.

RAW GREEN SALAD.

PREP TIME: 20 MINUTES
SERVES: 2

½ head broccoli (150g)
½ cucumber (150g)
70g kale
150g frozen peas
1 tsp fresh ginger, grated (15g)
1 garlic clove, grated
30g whole-grain mustard
small handful of fresh mint,
  chopped
40ml sesame oil
zest and juice of 1 lime
90g salted peanuts

SWAPS
FLAVOUR
• No broccoli? Use cauliflower,
carrots, peppers, green beans.
• No cucumber? Use fennel,
avocado, beetroot.
• No peas? Use edamame
beans, sweetcorn, broad beans.
• No kale? Use spinach,
Brussels sprouts, cabbage.

Remember when we told you this cookbook has no rules? Well, this 'raw' salad uses frozen peas - not technically correct, but an incredible addition nonetheless. We've also given you lots of swap suggestions that make the salad a different colour... Nearly-raw, sort-of-green salad, anyone?

1. Chop up every part of the broccoli. Cut the florets into bite-sized pieces and chop the stem into thin matchsticks.
2. Dice the cucumber and shred the kale with a knife.
3. Run cold water over the frozen peas to thaw them.
4. In a bowl, mix the ginger and garlic with the mustard, mint, sesame oil and the lime zest and juice to make the dressing.
5. Toast the peanuts in a dry pan and roughly chop.
6. Mix all the salad ingredients together and stir through the dressing. Then plate up and add a scattering of chopped peanuts on top.

VE

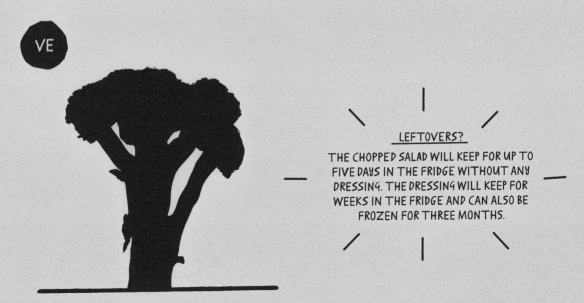

LEFTOVERS?
THE CHOPPED SALAD WILL KEEP FOR UP TO FIVE DAYS IN THE FRIDGE WITHOUT ANY DRESSING. THE DRESSING WILL KEEP FOR WEEKS IN THE FRIDGE AND CAN ALSO BE FROZEN FOR THREE MONTHS.

GRATED VEG FRITTERS.

PREP TIME: 15 MINUTES
COOK TIME: 10 MINUTES
SERVES: 2

120g plain flour
2 tsp mild curry powder
2 eggs
1 tsp baking powder
40ml any milk
1 courgette (180g)
1 carrot (75g)
1 garlic clove
3 spring onions, chopped (80g)
20ml olive oil
salt and pepper

FOR THE YOGHURT
80g any yoghurt
1 garlic clove, grated

SWAPS

FLAVOUR
• No carrot? Use sweetcorn, sliced spinach, mushrooms.
STRUCTURE
• No courgette? Use beetroot, chopped broccoli or cauliflower, shredded Brussels sprouts.

What could be 'grater' than super-simple, speedy fritters that help use up any veg you have in the fridge? Just make sure you grate everything into similar-sized chunks so the fritters cook evenly. And if you want to add extra spices like paprika, cumin or coriander, go wild – we dare you.

1. In a bowl, mix the flour, curry powder, eggs, baking powder and milk and stir to a smooth batter.
2. Grate the courgette, carrot and one garlic clove into the bowl and mix through the spring onions.
3. Heat a little oil in a frying pan and spoon in the batter into small mounds.
4. Cook for 1–2 minutes on a medium heat until nice and crispy before flipping. Flip a couple of times to make sure they're cooked evenly and all the way through.
5. Meanwhile, mix the yoghurt with the other clove of garlic (grated) and add a twist of black pepper.
6. When the fritters are ready, stack them up and serve with the garlicky yoghurt and a crisp, leafy salad. If you have any pickles kicking about, get them involved as well.

LEFTOVERS?
COOK A BIG BATCH OF FRITTERS TO STORE
IN THE FRIDGE FOR UP TO THREE DAYS,
OR IN THE FREEZER FOR THREE MONTHS.
SEPARATE THEM WITH BAKING PAPER SO
THEY DON'T STICK TOGETHER.

HERBY PEARL BARLEY SALAD.

PREP TIME: 20 MINUTES
COOK TIME: 45 MINUTES
SERVES: 2

70g pearl barley, or any grain of
   your choice
2 red onions (300g)
½ butternut squash (500g)
50ml olive oil
1 tsp ground cumin
40ml red wine vinegar
1 garlic clove, grated
2 tsp dried oregano
big handful of mint, parsley and
   chives, or any soft herbs
100g feta cheese, or vegan
   alternative
toasted nuts (optional)
salt and pepper

SWAPS

FLAVOUR
• No red onions? Use white
onions, fennel, leeks or a
courgette cut in half.
• No butternut squash? Use
sweet potatoes, pumpkin,
carrots, cauliflower.

This sunny salad has a carnival of textures and flavours in every bite - juicy pearl barley, rich, roasted butternut squash and tangy red onions with feta cheese. To make sure the pearl barley soaks up all the flavours, cook it in salted water and add the dressing while it's still warm. And feel free to toss in any other roasted veg you have - the more the merrier.

1. Preheat the oven to 200°C/180°C fan/400°F/gas 6.
2. Drop your pearl barley into a pan of salted water, bring to the boil and simmer for 45 minutes until tender.
3. Meanwhile, peel the red onions, keeping the root intact, and cut each one into four wedges. Place on a roasting tray.
4. Peel the squash and cut into large chunks then add to the onion tray. Drizzle all the veg with 20ml of the olive oil, cumin and salt and pepper and roast in the oven for 20-25 minutes.
5. When the pearl barley is cooked, drain it, then pour over the wine vinegar, the rest of the olive oil, the garlic and the oregano while it's still warm. Give everything a good mix and allow to cool. The pearl barley will soak up all the liquid so don't worry if it looks wet.
6. When the roasted vegetables are cooked, mix them with the pearl barley along with roughly-chopped soft herbs (any will do) and crumbled feta.
7. Toasted nuts would be a great final flourish, so sprinkle over any you have in the cupboard - or even toasted butternut squash seeds.

TOAST THE
SEEDS FROM THE
BUTTERNUT SQUASH
AS A SNACK OR SALAD
TOPPER – SEE PAGE 186.

LEFTOVERS?

THE SALAD WILL KEEP IN THE
FRIDGE FOR UP TO THREE DAYS
AND IS DELICIOUS WARM OR
COLD. ANY EXTRA ROASTED VEG
CAN BE FROZEN SEPARATELY.

SPICY VEG-STUFFED PEANUT WRAPS.

PREP TIME: 15 MINUTES
COOK TIME: 10 MINUTES
SERVES: 2

70g smooth peanut butter
30ml sriracha or other hot
  sauce
20ml white wine vinegar
50ml hot water
¼ red cabbage (120g)
2 carrots (150g)
2 large tortilla wraps
big handful of rocket
1 tin any beans, drained (225g)
handful of fresh coriander
120g Cheddar cheese or vegan
  alternative, grated
olive oil, for brushing

SWAPS

FLAVOUR
• **No cabbage? Use fennel,
cauliflower, peppers, red onions,
tomatoes, avocado.**
• **No carrots? Use courgettes,
beetroot.**
• **No rocket? Use spinach,
salad leaves, lettuce.**

> **LEFTOVERS?**
> USE ANY LEFTOVER
> CABBAGE AND CARROT
> IN THE RAW ZINGY SLAW
> (PAGE 132). THE PEANUT
> SAUCE ALSO GOES NICELY
> WITH THE VEGGIE KEBABS
> (PAGE 87).

Lots of veg starting to flop in the fridge? We think this is the best work-from-home, use-it-all-up, cram-it-all-in, should've-made-more, no-I'm-not-sharing lunch in the world. And that's a wrap.

1. Preheat the oven to 180°C/160°C fan/350°F/gas 4.
2. In a bowl, mix the peanut butter, hot sauce, vinegar and a splash of hot water to make a smooth creamy sauce.
3. Chop the cabbage into small, bite-sized pieces or strips and grate the carrots. Toss together.
4. Warm the tortilla wraps in a pan for 10–15 seconds, one at a time, so they are soft enough to roll.
5. Load up the wraps with the carrot and cabbage, rocket, beans, coriander, grated cheese and half the peanut sauce.
6. Fold over the wraps, tuck in the sides and roll up into a burrito shape.
7. Brush the wraps with a little oil, pop them in an ovenproof frying pan on a medium heat and fry for 2–3 minutes on both sides until nice and crispy. Slide into the oven for 5–10 minutes or until hot through.
8. Serve the wraps with more of the peanut sauce and enjoy.

MANY-VEG MINESTRONE.

PREP TIME: 20 MINUTES
COOK TIME: 30 MINUTES
SERVES: 4

2 white onions, diced (300g)
3 celery sticks, diced (120g)
2 carrots, diced (150g)
2 courgettes, diced (320g)
oil, for frying
2 garlic cloves, sliced
2 tins tomatoes (800g)
2 tins any beans (450g),
  drained
60g tomato purée
3 tsp smoked paprika
150g kale, chopped or shredded
600ml veg stock
salt and pepper

SWAPS

FLAVOUR
• No onion, celery, carrots or
courgettes? Use fennel, leeks,
celeriac, potatoes, squash,
green beans, mushrooms,
parsnips, peppers.
• No kale? Use, cabbage, cavolo
nero, broccoli.

Three simple steps; a billion possible ingredients. The reason
we love minestrone is that there are no limits to what you can
throw in - try it with whichever veg is in season and add pasta
or rice to bulk it out a bit more. Best of all, the flavour just
gets better over time, so you can make a big batch, freeze it in
bags and slurp for months to come.

1. In a heavy-bottomed saucepan, gently fry the onions, celery,
carrots and courgettes in a little oil for 5-10 minutes until nice
and soft.
2. Add the garlic and gently fry for 1-2 minutes before adding
all the remaining ingredients. Bring to the boil then simmer
for 20-30 minutes to allow all the flavours to combine. Taste,
and season with salt, pepper and any extra spice.
3. Serve with a big wedge of crusty bread and little grated
Parmesan, if you have any.

LEFTOVERS?
THE SOUP WILL KEEP IN THE
FRIDGE FOR UP TO FIVE DAYS,
OR IN THE FREEZER FOR
THREE MONTHS.

VEGGIE KEBABS WITH SATAY SAUCE.

PREP TIME: 20 MINUTES
COOK TIME: 40 MINUTES
SERVES: 2 (4 X 18CM
SKEWERS)

FOR THE SATAY SAUCE
80g smooth peanut butter
½ tin coconut milk (150ml)
20ml soy sauce
1 lime

FOR THE KEBABS
1 corn on the cob
2 red or yellow peppers (300g)
2 red onions (300g)
2 courgettes (320g)
big handful of salad leaves, to
   serve

SWAPS

STRUCTURE
• No peppers, onions or
courgettes? Use aubergine,
mushrooms, fennel, broccoli.
• No corn on the cob? Use
Brussels sprouts, cauliflower.

Dip, mix, stir, drizzle, fry. The great thing about satay sauce is that you can do almost anything with it - here, it's used as both a marinade and a dipping sauce. The kebabs are the perfect dish to prep the night before a BBQ or party - feel free to skewer any veg you need to use up.

1. Preheat the oven to 180°C/160°C fan/350°F/gas 4. If you're using wooden skewers, soak them in water for 15 minutes.
2. To make the satay sauce, pop the peanut butter, coconut milk, soy sauce and a good squeeze of lime juice into a saucepan and gently simmer. As the mixture warms up, the peanut butter will mix with the other ingredients.
3. Pour half the sauce into a bowl and the other half into a pot for dipping.
4. Carefully chop the corn on the cob into eight discs, blanch them in boiling water for 2-3 minutes until they're just softening, then leave them to cool.
5. Chop the peppers, onions and courgettes into big chunks and put them into the satay bowl along with the corn on the cob. Give everything a good mix.
6. Skewer your veg in any order you fancy, scraping any remaining dressing out of the bowl and smearing it on top.
7. Lay the skewers on a roasting tray and cook in the oven for 20-30 minutes, carefully turning them once or twice.
8. Serve the skewers with a fresh salad, or bulk them out with a bowl of Use-it-up fried rice (page 118) and your leftover dip.

LEFTOVERS?
THE VEG CAN BE PREPPED AND MARINATED IN THE
SATAY SAUCE AND WILL BE FINE IN THE FRIDGE FOR
UP TO THREE DAYS BEFORE COOKING. THE SAUCE
WILL KEEP IN THE FRIDGE FOR UP TO FIVE DAYS
AND FREEZE WELL FOR MONTHS.

SWEET POTATO & COCONUT SOUP.

PREP TIME: 20 MINUTES
COOK TIME: 40 MINUTES
SERVES: 4

50ml olive or sesame oil
thumb-size piece of ginger,
    grated (60g)
2 white onions (300g), diced
1 red chilli, roughly chopped
    with seeds in (remove for less
    heat)
3 sweet potatoes (480g)
4 garlic cloves, sliced
2 tsp smoked paprika
1 tin coconut milk (400ml)
400ml veg stock
zest and juice of 1 lime
400g spinach
big handful of fresh coriander
salt and pepper

SWAPS

FLAVOUR
• No sweet potatoes? Use
peeled butternut squash or
pumpkin, whole cauliflower.
• No spinach? Use broccoli
florets, shredded Brussels
sprouts, whole green beans.

Martyn says: 'I used to make buckets of this silky soup when I first learnt what a sweet potato was. It's super simple to chuck together – if you already have some roasted sweet potatoes, or have leftover mash, slide it into the soup too as you can hide all sorts in there. Oh and don't be afraid to pop in more lime, as it really lifts the flavour.'

1. Put the oil in a big saucepan and gently fry the ginger, onion and chilli over a low heat for 5 minutes until nice and soft.
2. Peel and chop the sweet potatoes, then add the chunks to the pan with the sliced garlic and paprika and mix well. Fry for 2-3 minutes.
3. Pour in the coconut milk and veg stock and bring to the boil, then reduce to a simmer and cook for 15-20 minutes or until the sweet potatoes are soft.
4. Use a hand blender to blitz to a smooth consistency, or a masher if you like chunkier soup.
5. Add the lime zest and juice, then season with salt and pepper.
6. Leave the soup simmering on a low heat and when you're nearly ready to serve, wilt in the spinach and stir through the fresh coriander. Serve with a big chunk of crusty bread.

LEFTOVERS?

THE SOUP WILL KEEP IN THE FRIDGE FOR UP TO THREE DAYS OR IN THE FREEZER FOR THREE MONTHS. IF YOU'RE PLANNING TO FREEZE IT, LEAVE OUT THE SPINACH AND ADD IT WHEN REHEATING.

VE

ODDBOX SOUP.

The rules for Oddbox soup? There are no rules. Make it red. Make it green. Make it silky smooth or fork-friendly. But whatever you do, don't just chuck it all in the pan and hope for the best – follow these basic steps to amp up the flavour, then go forth and blend.

1. START WITH A BASE

The secret to a good soup? Layers. A classic base is chopped carrots, onions and celery. But don't let that hold you back – other flavour-building veg include shallots, leeks, garlic or parsnips. Whatever you use, start by cooking it slowly in butter or oil in a big pan for 10–15 minutes.

2. COOK YOUR VEG

Either choose a hero veg or just chop up what you have – including leftovers from other recipes. To really bring out the flavours, try roasting your veg in the oven with oil and herbs before adding it to the pan, or chargrilling it. Or just sauté it with a little oil before adding liquid.

3. ADD SEASONING

What will add oomph? Go Moroccan-style with cinnamon, cumin and coriander. For Thai flavours, add chilli, lemongrass, ginger and coriander. For Spanish or Moroccan vibes, add a pinch of smoked paprika. Or for a classic veg soup, add mixed dry herbs. There are no limits here, so add a sprinkling of whatever you fancy, plus salt and pepper.

4. POUR IN LIQUID

Add a liquid of your choice – see page 185 for our Scrappy stock recipe, or try coconut milk. Don't just use water, though, or your soup won't taste of much. Whatever you go for, you'll need enough to completely cover the vegetables. Leave everything to simmer until the veg is soft.

5. BLITZ – OR DON'T

Like your soups smooth? Give everything a good blitz with a hand blender. Or separate half the mixture before blitzing, then pour it back in for a chunkier number. You can even add pasta, beans or grains to bulk it up more.

6. ADD A FINAL FLOURISH

A dollop of yoghurt? Toasted seeds (page 186)? Fresh herbs? Croutons? Truffle oil? A sprinkling of Parmesan? A handful of fresh spinach? It's up to you. Add whatever you fancy, and soup is served.

MAINS.

FROM SPEEDY MID-WEEK
DINNERS TO CREATIVE
WEEKEND SPECTACULARS.

107

118

120

102

124

116

WONKY VEGGIE BURGER.

PREP TIME: 25 MINUTES
COOK TIME: 15 MINUTES
SERVES: 4

FOR THE BURGER

1 tin red kidney beans (225g),
    drained
1 white onion, diced (150g)
1 carrot, grated (75g)
1 small tin sweetcorn (100g),
    drained
oil, for frying
1 garlic clove, grated
2 tsp paprika
2 tsp dried oregano
½ tsp ground cumin
1 egg
40g breadcrumbs
small handful of fresh coriander
salt and pepper

FOR THE BUN AND TOPPINGS

½ garlic clove, crushed
50g any yoghurt
2 burger buns
1 avocado, sliced
any garnish (try sliced red
    onion, chopped lettuce, sliced
    tomato, grated cheese)

SWAPS

FLAVOUR
• No onions? Use leeks, sliced
fennel or spring onions.
• No carrots? Use finely diced
aubergine, blitzed mushrooms,
sliced peppers.

One, it's a burger. Two, you can be creative with the veg mix *and* your toppings. Three, we get to use the word patty, which is almost as much of a treat as having a burger. There's a whole lot to love about this recipe – Martyn's given it a Mexican vibe with sweetcorn and spice, but feel free to make it your way. You can even swap the mashed beans for roasted root veg. Fancy.

1. Preheat the oven to 180°C/160°C fan/350°F/gas 4.
2. To make the patties, spread out the beans on a baking tray. Pop in the oven for 10 minutes so they start to dry out – don't worry if they split.
3. Meanwhile, gently fry the onion, carrot and sweetcorn in a little oil for 5 minutes, so everything starts to caramelise.
4. Add the grated garlic and all the spices, then fry for 1-2 minutes.
5. Mash half the beans in a bowl. Then add the onion mixture, the remaining beans, egg, breadcrumbs and fresh coriander.
6. Fry a little pinch in the pan, taste and adjust the seasoning if needed.
7. Shape the patty mixture to the size of your buns, and fry the patties in a little oil for 1-2 minutes on both sides until nice and crispy. Transfer to the oven and cook for 5 minutes or until hot through.
8. At the same time, mix the half garlic clove and the yoghurt, then season with salt and pepper.
9. Toast the buns, pop in the patties and add the sliced avocado and other garnishes with a dollop of garlic yoghurt. Delish.

LEFTOVERS?

THE PATTY MIX WILL KEEP IN THE FRIDGE FOR THREE DAYS OR IN THE FREEZER FOR THREE MONTHS. TO FREEZE, SHAPE THE PATTIES AND PLACE A PIECE OF BAKING PAPER BETWEEN EACH ONE SO THEY DON'T STICK TOGETHER.

ROOT-VEG ROSTIS.

PREP TIME: 20 MINUTES
COOK TIME: 5 MINUTES
SERVES: 2

2 floury potatoes (300g)
1 parsnip (120g)
5 eggs
20g flour
pinch of nutmeg
2 knobs of butter
olive oil, for frying
glug of white wine vinegar
½ white cabbage (300g),
  shredded
salt and pepper

SWAPS

FLAVOUR
• No cabbage? Use kale,
spinach, cavolo nero.
STRUCTURE
• No potatoes? Use sweet
potatoes.
• No parsnip? Use carrots,
celeriac, onions, fennel.

LEFTOVERS?

THE ROSTIS WILL KEEP
IN THE FRIDGE FOR UP TO
THREE DAYS. TO FREEZE,
SEPARATE EACH ONE WITH
A PIECE OF BAKING PAPER
AND STORE FOR THREE
MONTHS. THEN REHEAT
IN THE OVEN.

You might be wondering if this dish is in the right section –
but why should breakfast have all the fried fun? These rostis
are a quick and delicious way to use up potatoes and parsnips,
and you can add other root veg for the ride. Eggs and spinach
on top. Delicious weekday dinner – done.

1. Grate the potatoes and parsnip, skin-on, onto a clean tea
towel and squeeze out as much water as possible.
2. Pop the grated roots in a bowl and mix them with one egg
and the flour. Season with salt, pepper and a pinch of nutmeg.
3. Roughly shape the rosti mix into patties.
4. Melt a good knob of butter in a frying pan over
medium heat with a little dribble of oil, and fry the rostis for
3–5 minutes on both sides. (If they're very thick, you can also
put them into the oven at 180°C/160°C fan/350°F/gas 4 for
5 minutes or until they're cooked through.)
5. Meanwhile, bring a deep saucepan of water to the boil, add
the vinegar and reduce the heat to a gentle simmer.
6. Crack in the eggs and poach for 3–3½ minutes, then
remove them with a slotted spoon and place on a paper towel.
7. Melt another big knob of butter in the frying pan you used
for the rostis and add the shredded cabbage. Add a splash of
water and cook for 3–5 minutes until softened, then season
with salt and pepper.
8. Pop a couple of rostis on a plate, top with buttered cabbage
and the poached eggs and get stuck in.

NO-LIMITS LASAGNE.

PREP TIME: 20 MINUTES
COOK TIME: 1½ HOURS
MINUTES
SERVES: 6

2 onions, chopped (300g)
2 sticks celery, diced (100g)
1 bulb fennel, diced (200g)
50ml olive oil
2 courgettes, cubed (300g)
4 garlic cloves, sliced
2 tins chopped tomatoes
  (800g)
big bunch of basil leaves,
  chopped
8-10 lasagne sheets
salt and pepper

FOR THE BECHAMEL
60g any butter
60g plain flour
600ml any milk
120g Cheddar cheese or vegan
  equivalent, plus extra for the
  top, grated
1 tsp Dijon mustard

SWAPS

FLAVOUR
• No onions, celery or fennel?
Use leeks, spring onions,
aubergine, mushrooms.
• No courgettes? Use peppers,
broccoli, cauliflower.

LEFTOVERS?
THE LASAGNE WILL KEEP
IN THE FRIDGE FOR UP
TO THREE DAYS OR IN
THE FREEZER FOR THREE
MONTHS. DIVIDE INTO
INDIVIDUAL PORTIONS
BEFORE FREEZING TO MAKE
IT EASIER TO DEFROST.

A bit like soup, lasagne is one of those dishes you can really take off-recipe. Layer up leftover roasted veg or any purées you've made. Make a sauce out of offcuts and knobbly bits (see page 191). And keep portions in the freezer for a delicious dinner you can whip out after a long day.

1. Preheat the oven to 180°C/160°C fan/350°F/gas 4.
2. Start by making a tomato sauce. In a large saucepan over a medium heat, fry the onions, celery and fennel in olive oil for 5-8 minutes until softened and caramelised.
3. Add the courgettes, garlic and chopped tomatoes. Fill the tins with a little water to rinse them out and pour the water into the pan, then bring the mixture to a boil.
4. Simmer for 30-40 minutes, or until the sauce is nice and thick. Stir through the basil and season to taste with salt and pepper.
5. While the sauce is bubbling away, make the bechamel. Melt the butter in a saucepan and add the flour, then cook gently for 1 minute until the flour starts to froth a little.
6. Slowly add the milk, whisking as you pour. Then gently simmer the sauce for a couple of minutes, whisking until it thickens.
7. Mix in the grated cheese and Dijon mustard, then season to taste with salt and pepper.
8. Layer the lasagne in an ovenproof dish. Start with the tomato sauce, cover it with pasta sheets then bechamel and repeat until both sauces are used up. Finish with bechamel and more cheese.
9. Put the dish on a roasting tray to catch any drips and bake in the oven for 30-45 minutes, or until bubbling and golden brown.
10. Serve the lasagne on its own, or with a green salad and garlic bread.

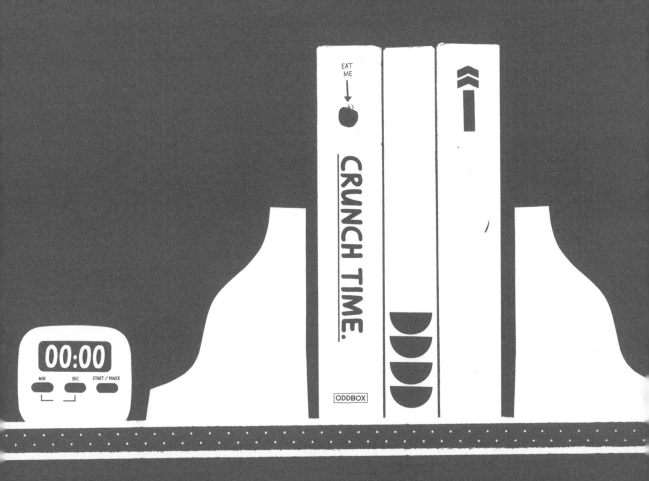

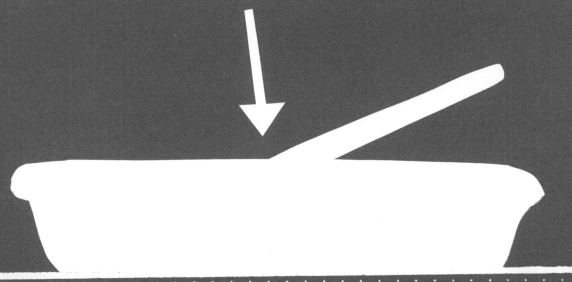

BUBBLING LASAGNE
GOODNESS

SWEET POTATO FALAFELS IN PITTA.

PREP TIME: 20 MINUTES
COOK TIME: 30 MINUTES
SERVES: 2

3-4 sweet potatoes (320g)
20ml olive oil
2 tsp smoked paprika
1 tsp ground cumin
120g quinoa
1 tbsp plain flour
1 tin chickpeas, drained (225g)
3 garlic cloves, grated
handful of fresh coriander,
  chopped
50g red cabbage
½ cucumber (100g)
20g tahini
juice of 1 lemon
50ml water
2 pitta breads

SWAPS

FLAVOUR
• No cucumber or cabbage?
Use lettuce, avocado, fennel,
red onions, tomatoes, radishes.
STRUCTURE
• No sweet potatoes? Use
butternut squash, parsnips,
potatoes.

MARTYN'S TOP TIP:
AS THE FALAFELS ARE
DELICATE, TRY SLIDING
THEM UP THE SIDES OF
THE PAN SO THEY ROLL
OVER, INSTEAD OF LIFTING
THEM OUT TO FLIP.

All hail the most flexible falafels in the land. (Try saying that with a mouthful.) We've used sweet potatoes in the recipe, but you could also use squash, potatoes and even parsnips. And there's no need to buy a red cabbage just to fill your pitta pockets - any crunchy salad will go down a treat. Finally, don't forget to make a few extra for the freezer - you'll thank yourself on a rainy day.

1. Preheat the oven to 180°C/160°C fan/350°F/gas 4.
2. Chop the sweet potatoes, leaving the skin on, into eight even-sized pieces. Pop them into a bowl and drizzle with some of the oil, paprika and cumin and mix well.
3. Lay the pieces on a baking tray and roast for 20-30 minutes, or until the potatoes are nicely browned.
4. Meanwhile, rinse the quinoa under cold water for 30 seconds, pop into a saucepan and cover with double the quantity of water. Simmer with the lid off until the water has evaporated, then turn off the heat, put on the lid and leave to steam.
5. When the sweet potatoes are cooked, add the chickpeas and quinoa to the baking tray and mash everything together, stir through the flour and mix well so it has a consistency that will hold a shape.
6. Add two-thirds of the garlic and the coriander, season to taste and shape the mixture into mini patties or balls.
7. Warm a non-stick pan on a medium heat and drizzle in a little oil. Then gently fry your falafels for 2-3 minutes on both sides, until crispy all over. Be careful as they're quite delicate - don't worry if little bits of quinoa fall off.
8. Prepare the other fillings by shredding or dicing the cabbage and thinly slicing the cucumber.
9. Mix the tahini, remaining garlic and the lemon juice in a bowl. Add water slowly so it doesn't get too runny, and keep mixing until you have a thick dressing. Season to taste.
10. Toast the pittas and stuff them with the cabbage and cucumber, falafels and sauce. Yum.

BRAISED FENNEL WITH BEAN PURÉE.

PREP TIME: 15 MINUTES
COOK TIME: 30 MINUTES
SERVES: 2

FOR THE FENNEL
2 fennel bulbs (500g)
30ml olive oil
zest and juice of 1 lemon
salt and pepper

FOR THE PURÉE
2 garlic cloves
sprinkle of nutmeg
100ml any milk
1 tin any beans, drained (225g):
    butter, pinto or haricot
    work well

FOR THE DRESSING
30g hazelnuts (blanched and
    peeled) or almonds or
    pine nuts
big handful of parsley
10ml olive oil
20g whole-grain mustard
20g capers
salt and pepper

SWAPS

FLAVOUR
• No fennel? Use hispi cabbage,
new potatoes, squash, beetroot,
halved leeks.
• No parsley? Use dill,
tarragon, chives, basil.

Bring out the mellow side of fennel with a creamy, garlicky bean purée. In fact, bring out the mellow side of any vegetable – with a tin or two of beans in the cupboard, you'll always be able to whip up a no-fuss purée that tastes much fancier than it is. (Swiping it across the plate like a chef: optional.)

1. Preheat the oven to 180°C/160°C fan/350°F/gas 4.
2. Lay a long piece of baking paper on a baking tray. Make it big enough to allow you to fold it back on itself.
3. Cut the fennel bulbs in half and then in half again, so you have eight wedges in total. Drizzle with oil, lemon zest and juice, then season with salt and pepper.
4. Lay the pieces on the baking tray, fold the baking paper over them, then bake in the oven for 30 minutes.
5. Meanwhile, grate the garlic and a little nutmeg into the milk. Gently warm them in a pan with the beans, simmering for 3–5 minutes.
6. Use a blender to blitz the bean mixture until it has a slightly loose, mash-like consistency.
7. Spoon the purée back into the saucepan and warm through. Season to taste and add a little more milk if it's too thick.
8. To make the dressing, roughly chop the nuts and toast in a pan for 2–3 minutes.
9. Chop the parsley and mix with the oil, mustard and capers. When the hazelnuts are toasted, pop them into the dressing and season if needed (it should be quite oily).
10. Spoon the purée on to a plate with the braised fennel and finish with the herby dressing.

LEFTOVERS?

THE FENNEL WILL KEEP IN THE FRIDGE FOR UP TO THREE DAYS. OR YOU CAN USE IT AS THE BASE OF A SOUP, OR ADD IT TO THE HERBY PEARL BARLEY SALAD ON PAGE 84.

SMOKY AUBERGINE LENTILS.

PREP TIME: 15 MINUTES
COOK TIME: 45 MINUTES
SERVES: 2

2 aubergines (750g)
5 tomatoes (400g)
80g puy lentils
½ celeriac (400g)
30ml olive oil
1 onion, thinly sliced (150g)
3 garlic cloves, sliced
50g any butter
2 tsp curry powder
1 tsp smoked paprika
dollop of any yoghurt, to serve
handful of any fresh herbs
salt and pepper

SWAPS

FLAVOUR
• No aubergine? Use blackened
peppers (same technique). Or
halved beetroot or carrots.
• No celeriac? Use potatoes,
sweet potatoes, squash,
parsnips.

LEFTOVERS?

KEEP FOR UP TO THREE
DAYS IN THE FRIDGE OR
IN THE FREEZER FOR
THREE MONTHS.

This deliciously chunky dish is perfect for cooking in bulk, and as you only need half a celeriac, you can roast the other half at the same time to use in salads, soups or stews. Two meals at once? Now you're talking.

1. Preheat the oven to 200°C/180°C fan/400°F/gas 6.
2. Prick the aubergines all over with a fork and lay them in a hot chargrill pan to blacken the skin all over. Be patient – they'll take about 5 minutes on each side.
3. Slice the tomatoes in half and spread out on a baking tray with the blackened aubergines. Slide the tray into the oven for 15-20 minutes until the aubergine and tomatoes are soft and caramelised.
4. Meanwhile, pop the lentils in a pan of water, bring to the boil and simmer for 20-30 minutes or until tender. Drain and set to one side.
5. Wash the celeriac to remove any mud and chop into small cubes.
6. In a large saucepan, fry the cubes in oil over medium heat for 5-8 minutes, until caramelised and just turning soft.
7. Turn down the heat and add the onion, garlic, butter, curry powder and paprika. Gently fry for 2-3 minutes.
8. Take the aubergine and tomatoes out of the oven. Hold on to the aubergine stalks and roughly chop the flesh, then add to the cooked lentils with the tomatoes.
9. Add a splash of water and bring to the boil, then add any fresh herbs you have and salt and pepper if needed.
10. Serve with a big dollop of yoghurt.

LEFTOVERS?
KEEP THE TOMATO SAUCE IN THE FRIDGE FOR UP TO FIVE DAYS, OR FREEZE FOR THREE MONTHS. STORE THE FILLING SEPARATELY IN THE FRIDGE.

STUFFED PASTA SHELLS.

PREP TIME: 30 MINUTES
COOK TIME: 50 MINUTES
SERVES: 2

8 medium tomatoes (450g)
60ml olive oil
2 tbsp oregano
3 garlic cloves, sliced
splash of red wine vinegar
   (optional)
sprinkling of sugar (optional)
200g spinach, roughly chopped
130g ricotta, or feta, goat's
   cheese, Cheddar, vegan
   alternative
20g Parmesan, grated, or vegan
   alternative
16 conchiglioni rigati (large
   pasta shells), or use pasta
   sheets or cannelloni
a few basil leaves, to serve
salt and pepper

SWAPS

FLAVOUR
• No spinach? Use shredded
kale or cavolo nero, grated
courgette, defrosted peas.
STRUCTURE
• No tomatoes? Use tinned
tomatoes.

She stuffs sea shells by the seashore. This is a perfect dish for making in bulk – keep the rich, roasted tomato sauce in the freezer for quick dinners, and feel free to switch up the ricotta filling with butternut squash, feta or goat's cheese. If you've made the Odds & ends sauce on page 191, you can also use it here.

1. Preheat the oven to 180°C/160°C fan/350°F/gas 4.
2. Slice the tomatoes in half. Then drizzle with half the oil, season with oregano and salt and pepper and roast in the oven for 20–30 minutes, or until nicely caramelised.
3. Meanwhile, put a saucepan on a very low heat and fry the sliced garlic in the remaining oil for 1–2 minutes.
4. Add the roasted tomatoes and mash or blitz to a sauce. Simmer for 10 minutes. The sauce should be nice and liquid as it will thicken up in the oven – add a splash of water if it's too dry.
5. Taste the sauce and season with salt and pepper. Add a dash of vinegar and sugar if the tomatoes are a bit tart.
6. Pop the spinach into a frying pan and gently fry with a splash of water to wilt. Drain in a colander and give it a squidge to remove any liquid.
7. Put the spinach in a bowl with the cheeses and mash everything together well. Season to taste with salt and pepper.
8. At the same time, cook the pasta in salted boiling water. Drain and run under cold water so the shells are cool enough to handle.
9. Pour the tomato sauce into a baking dish, fill the shells with the spinach mixture and nestle them into the sauce.
10. Turn the oven up to 200°C/180°C fan/400°F/gas 6 and bake for 15–20 minutes, or until the sauce is bubbling. Then sprinkle over a little more Parmesan and a few basil leaves. Serve with a leafy salad.

COCONUT CASHEW CURRY.

PREP TIME: 15 MINUTES
COOK TIME: 30 MINUTES
SERVES: 4

100g cashew nuts
1 cauliflower (450g)
30ml olive oil
2 tsp cumin seeds
2 tsp coriander seeds
1 onion (150g)
3 garlic cloves
thumb-size piece of ginger
  (50g)
60ml vegetable oil
1 stick lemongrass
2 tsp mild curry powder
2 tins coconut milk (800ml)
1 tsp sugar
juice of 1 lemon
1 tin chickpeas, drained (250g)
salt and pepper

SWAPS

FLAVOUR
• No cauliflower? Use
butternut squash, broccoli,
aubergine, carrots, parsnips.
• No cauliflower leaves? Use
spinach, green beans, kale,
cabbage, pak choi.

MARTYN'S TOP TIP:
FREEZE THE LEMONGRASS
AND GINGER AND THEN
GRATE RATHER THAN
CHOP IT. MUCH EASIER!

Cauliflower. Cashews. Coconut milk. Despite what the ingredients list may suggest, this deliciously versatile curry will work with pretty much any veg – whichever letter it begins with. You can even add a dollop of nut butter instead of cashews to create the smooth texture. Totally dreamy (or should that be creamy?).

1. Preheat the oven to 200°C/180°C fan/400°F/gas 6 and put the cashews in a bowl of water to soak.
2. Break the cauliflower into bite-sized pieces, saving the stalk and leaves. Put the pieces on a tray, drizzle with olive oil and season with salt and pepper. Then roast in the oven for 15 minutes or until tender.
3. Meanwhile, toast your cumin and coriander seeds in a pan for 1–2 minutes until they start to pop, then blitz or grind them to a powder.
4. Peel the onion and garlic, and blitz them to a paste with the ginger in a food processor.
5. Put the vegetable oil in a cold frying pan, spoon in the onion paste and gently fry for 3–5 minutes until it starts to go translucent. Chop or grate the lemongrass and add to the pan.
6. Add the ground spices and curry powder and cook for 1–2 minutes. Dial up the spices if you like hotter curries, or add chopped fresh chilli.
7. Drain the soaked cashews and blitz with the coconut milk into a smooth liquid, then add to the pan and gently simmer for ten minutes.
8. Taste the sauce and season with salt, sugar and lemon juice.
9. Chop up the cauliflower leaves and add them to the sauce with the thinly sliced stalk, roasted florets and chickpeas.
10. Cook for a few minutes until the leaves and stalk soften, then serve with rice.

LEFTOVERS?
THE CURRY WILL KEEP FOR UP TO FIVE
DAYS IN THE FRIDGE, OR FOR THREE
MONTHS IN THE FREEZER.

DON'T WANT TO INCLUDE THE
CAULIFLOWER LEAVES AND
STALK? SAVE THEM TO MAKE A
PICKLE – SEE PAGE 190.

VEGGIE COTTAGE PIE.

PREP TIME: 20 MINUTES
COOK TIME: 1 HOUR
SERVES: 4-6

1 celeriac (1kg)
50ml olive oil
150g puy lentils (dry)
1 onion, diced (150g)
2 carrots, diced (150g)
8 mushrooms (300g)
50g tomato purée
2 tins tomatoes (800g)
3 garlic cloves, sliced
200ml veg stock
50g butter, olive oil or vegan
  cheese
salt and pepper

SWAPS

FLAVOUR
• No mushrooms? Use chopped
asparagus, diced aubergine,
sliced peppers, leeks, fennel.
• No carrots? Use sliced leeks,
fennel, celery.
STRUCTURE
• No celeriac? Use potatoes,
parsnips, cauliflower, swede,
sweet potatoes, peeled
butternut squash.

> ### LEFTOVERS?
> KEEP IN THE FRIDGE
> FOR UP TO FIVE DAYS,
> OR IN THE FREEZER FOR
> THREE MONTHS.

This is a lovely chunky pie, perfect for cooking in bulk. The filling is also delicious on its own, dolloped on a jacket potato or flatbread, or with a leafy salad. To make it vegan, replace the butter in the mash with olive oil or a plant-based cheese. And if you want to use precooked lentils, just double the weight (red lentils or split peas will be too mushy). Easy as pie.

1. Preheat the oven to 180°C/160°C fan/350°F/gas 4.
2. Chop the celeriac into thick wedges, leaving the skin on. Place on a baking tray with a drizzle of oil, add a good pinch of salt and pop in the oven for 30 minutes, or until soft enough to mash.
3. Meanwhile, pop the lentils into a pan of water, bring to a boil, then simmer for 20-30 minutes until tender.
4. Heat a large saucepan over a medium heat. Add the rest of the olive oil and gently fry the diced onion and carrots for 3-5 minutes until just starting to turn soft.
5. Blitz or slice the mushrooms and add to the pan, then cook for 5-10 minutes until super-soft.
6. Add the tomato purée, tinned tomatoes, sliced garlic, stock and cooked lentils. Bring to the boil and let everything bubble away until nice and thick. Season to taste with salt and pepper.
7. When the celeriac is roasted, roughly mash with butter (you can add cheese or mustard if you like). Season to taste. Turn up the oven to 200°C/180°fan/400°F/gas 6.
8. Spoon the lentil and mushroom filling into a baking dish, and top with the celeriac mash.
9. Slide into the oven and bake for 20-30 minutes until golden and bubbling.

CREAMY ROOT BAKE.

PREP TIME: 20 MINUTES
COOK TIME: 40 MINUTES
SERVES: 4

600ml double cream or vegan
   alternative
200ml any milk
4 garlic cloves
6 sprigs fresh thyme
½ tsp grated nutmeg
½ celeriac (400g)
4 potatoes (400g)
100g any cheese (Grùyere
   would be great) or vegan
   cheddar, grated
salt and pepper

SWAPS

FLAVOUR
• No celeriac or potatoes?
Use sweet potatoes, parsnips,
turnips or peeled squash or
pumpkin. Also try adding sliced
Brussels sprouts, broccoli,
cauliflower, leeks.

From side dish to star of the show. This dreamy bake is based on a dauphinoise, but it's too special to play second fiddle. Experiment with different herbs in the creamy sauce. And add any veg you need to use up - spinach can be wilted down in the sauce, while shredded cabbage and Brussels sprouts can be layered with the potato and celeriac. Versatile, see?

1. Preheat the oven to 180°C/160°C fan/350°F/gas 4.
2. Gently warm the double cream and milk in a large saucepan. Grate in the garlic, shred and add the thyme leaves and season with nutmeg, salt and pepper.
3. Meanwhile, slice the celeriac and potatoes nice and thinly (leaving the skin on) with a mandolin or knife. (Be careful if using a mandolin.) Slices about 3-4mm thick are ideal.
4. Pop the sliced celeriac and potato into a large bowl and pour over the garlic cream, giving it a really good mix.
5. Layer the potato and celeriac in an ovenproof dish, making sure it's deep enough for all the liquid.
6. Pour the cream over the top, press down the layers and place the ovenproof dish on a baking tray in case the sauce spills over.
7. Bake in the oven for 30-40 minutes. With 10 minutes to go, sprinkle grated cheese over the top.
8. It's cooked when you can slide a knife in easily and the top is brown and bubbling. Serve with fresh sautéed greens and enjoy.

LEFTOVERS?

LEAVE THE DISH TO COOL,
THEN POP IT IN THE FRIDGE
OVERNIGHT. EITHER EAT
WITHIN THE NEXT THREE
DAYS, OR SLICE INTO INDIVIDUAL
PORTIONS TO FREEZE FOR
THREE MONTHS.

MADE A BIG BATCH?
SEE PAGE 134 FOR OUR
LEFTOVER RISOTTO
ARANCINI RECIPE.

ANY-VEG RISOTTO.

PREP TIME: 20 MINUTES
COOK TIME: 30 MINUTES
SERVES: 2

600ml veg stock
60ml olive oil
1 white onion, diced (150g)
3 garlic cloves, grated or sliced
180g arborio rice
1 leek, thinly sliced (200g)
1 courgette, diced (160g)
100g cream cheese or vegan
    alternative
150g defrosted garden peas
handful of fresh basil and
    parsley
zest and juice of 1 lemon
salt and pepper

SWAPS

FLAVOUR
• No onion? Use shallots,
fennel, leeks.
• No courgette or leek? Use
sliced mushrooms, fennel,
asparagus. Or finely diced
celeriac, parsnips, squash.
• No peas? Use broad beans,
rocket, baby spinach.

Lots of veg to use up? With a really tasty risotto base on the go, you're free to chop up and chuck in a little bit of anything you fancy. We've included leek and courgette on the ingredients list, but fennel, kale, squash and spinach work just as well. If you're batch-cooking, try taking out freezable portions while the rice is slightly undercooked. That way, when you reheat it with a glug of stock, you'll end up with perfectly cooked rice with a bit of bite.

1. Heat the stock in a pan and bring to the boil.
2. Pour a glug of olive oil into another large saucepan and gently fry the onion for 2–3 minutes. Then add the garlic and fry for a further 1–2 minutes.
3. Pour the rice into the pan with the onions and garlic and toss it around gently.
4. Add the stock to the rice pan a ladle at a time, mixing constantly until it's absorbed. Keep going until all the stock is gone.
5. At the same time, sauté the leek and courgette in another pan with a drizzle of oil and season with salt and pepper. Gently fry for 2–3 minutes or until the veg is tender and lightly caramelised.
6. Add the veg to the rice, then mix through the cream cheese, peas, herbs, lemon zest and juice. Give everything a good stir and cook for 1–2 minutes, to warm the peas through.
7. Serve with a salad and crusty bread.

LEFTOVERS?
KEEP ANY LEFTOVER
RISOTTO IN THE FRIDGE
FOR UP TO FIVE DAYS,
OR IN THE FREEZER FOR
THREE MONTHS.

MISO-GLAZED AUBERGINE.

PREP TIME: 20 MINUTES
COOK TIME: 30 MINUTES
SERVES: 2

60ml sesame oil
20g miso paste
40g honey or maple syrup
30ml soy sauce
small piece of fresh ginger,
   grated (15g)
2 garlic cloves, grated
1 fresh chilli, chopped (optional)
1 aubergine (350g)
120g basmati rice
3 pak choi (300g)
2 spring onions, sliced, to
   garnish
salt and pepper

SWAPS

FLAVOUR
• No pak choi? Use spinach,
kale, cabbage.
STRUCTURE
• No aubergine? Use
cauliflower or broccoli (florets
and stem), halved carrots or
courgettes, whole Brussels
sprouts or big mushrooms.

Who knew aubergine could taste this amazing? Drenched in sticky-sweet sauce, the wedges are just as delicious eaten cold as hot. Instead of pak choi, you could try our Tangy miso greens (see page 149) as an accompaniment. If you don't have rice, go for noodles, couscous or bao buns. And if you have any leftovers (unlikely, we admit), try tossing the aubergine through a crisp salad.

1. Preheat the oven to 200°C/180°C fan/400°F/gas 6.
2. Make the miso glaze. In a bowl, mix the sesame oil, miso paste, honey, soy sauce, grated ginger and garlic. Feel free to add fresh chilli if you like a bit of heat.
3. Cut the aubergine lengthwise into eight thick wedges and toss them in the glaze.
4. Lay the wedges on a lined baking tray and roast in the oven for 20-30 minutes, turning once or twice to baste with the glaze, until nicely caramelised and sticky.
5. Meanwhile, rinse the rice under cold water until the water runs clear. Put two parts cold water to one part rice in a pan - so for 120g of rice, add 240ml of water.
6. Add a good pinch of salt, pop on a lid and bring to the boil. As soon as the water boils, turn down to a simmer and cook for 8-10 minutes.
7. Turn off the heat, leave on the lid and allow the rice to steam for 5 minutes.
8. Cut your pak choi into quarters and steam or blanch in a pan of boiling water for 3 minutes, until tender.
9. Fluff the rice with a fork and scoop into a bowl. Top with the sticky aubergine and steamed pak choi and garnish with spring onions and any extra sauce.

LEFTOVERS?

IF YOU MAKE A BIGGER BATCH OF THE MISO GLAZE, YOU CAN ALSO USE IT AS A SALAD DRESSING IN THE RAW ZINGY SLAW (PAGE 132), THE RAW GREEN SALAD (PAGE 81) OR TANGY MISO GREENS (PAGE 149).

CHUNKY PESTO PASTA WITH TOMATOES.

PREP TIME: 15 MINUTES
COOK TIME: 3 HOURS
SERVES: 2

FOR THE OVEN-DRIED TOMATOES
325g cherry tomatoes
20ml olive oil
1 garlic clove, grated

FOR THE PESTO PASTA
50g pine nuts, or almonds, cashews, hazelnuts
small bunch of basil, chopped
1 garlic clove, grated
100g Parmesan cheese, grated, or vegan alternative
70ml olive oil
juice of 1 lemon
180g pasta
120g kale, chopped
300g tenderstem broccoli
salt and pepper

SWAPS

FLAVOUR
• No broccoli? Use discs of courgette or aubergine, sliced peppers, halved leeks or portobello mushrooms.
• No kale? Use spinach, cavolo nero, rocket.
STRUCTURE
• No cherry tomatoes? Use bigger tomatoes or sun-dried tomatoes.

True, this is a bit more work than stirring a jar of pesto through your pasta. But the results are 100% worth it, and the oven-dried cherry tomatoes are a great fridge staple for salads, sandwiches and antipasti boards. Make a big batch of tomatoes in advance if you can, then the rest of the recipe is super-quick.

1. To make the oven-dried cherry tomatoes, preheat the oven to 140°C/120°C fan/275°F/gas 1.
2. Slice the tomatoes in half, pop in a bowl and mix with the oil, garlic and a pinch of salt. Lay them on a roasting tray, cut side up, and slide into the oven for 2-3 hours. (A shorter roasting time gives you juicier tomatoes; longer will make them drier and chewy.) Turn after 30-40 minutes, then when they're ready, set to one side or keep in the fridge.
3. To make the pesto pasta, start by toasting the pine nuts and chopping them roughly with a knife.
4. In a bowl, mix the basil, pine nuts, garlic, Parmesan, olive oil and half the lemon juice.
5. Add the tomatoes and mix through.
6. Meanwhile, cook the pasta in salted boiling water. Drop the kale into the pan for the last 2 minutes, then drain everything, keeping a little pasta water for the sauce.
7. Drizzle the broccoli with oil and chargrill in a frying pan for a couple of minutes. Put it in a bowl and squeeze over the rest of the lemon juice.
8. To serve, mix the broccoli, pasta and pesto in a pan and warm over a low heat. Use a little pasta water to help loosen the sauce. Add salt and pepper if you like and get stuck in.

> ## LEFTOVERS?
> KEEP ANY LEFTOVER PASTA IN THE FRIDGE FOR UP TO THREE DAYS. THE CHERRY TOMATOES WILL KEEP FOR A COUPLE OF WEEKS IN THE FRIDGE IN AN AIRTIGHT CONTAINER – OR ADD THEM TO THE ROASTED-VEG BAGELS (PAGE 59) OR PEARL BARLEY SALAD (PAGE 84).

NUTTY SQUASH WITH LEMON YOGHURT.

PREP TIME: 20 MINUTES
COOK TIME: 5 MINUTES
SERVES: 2

1 tsp cumin seeds
1 tsp coriander seeds
½ tsp curry powder
1 butternut squash (1kg)
30ml olive oil
zest and juice of 1 lemon
100g any yoghurt
1 garlic clove, grated
40g walnuts or almonds,
  pine nuts, hazelnuts, pecans,
  cashews, roughly chopped, to
  serve
salt and pepper

SWAPS

STRUCTURE
• No butternut squash? Use
pumpkin, sweet potatoes,
potatoes, carrots, cauliflower.

Roasting squash is the best way to bring out the naturally sweet flavour, and to make the skin soft enough to eat. We like to throw this dish into the mix with other sharing plates - the spice mix is beautifully balanced with the tangy yoghurt, while the walnuts give a bit of crunch. Perfection.

1. Preheat the oven to 200°C/180°C fan/400°F/gas 6.
2. Gently toast the cumin and coriander seeds in a dry pan for 1-2 minutes until they start to pop.
3. Use a spice grinder or pestle and mortar to grind them into a coarse powder. Then mix through the curry powder and add a pinch of salt.
4. Carefully split the squash in half, scoop out the seeds then chop the squash into big, chunky wedges. Lay them in a baking tray, drizzle with oil and sprinkle over the spice mix.
5. Roast in the oven for 20-30 minutes or until the wedges are nicely caramelised.
6. Meanwhile, zest the lemon and mix with the yoghurt. Stir through the lemon juice and grated garlic, and season with salt and pepper.
7. Dollop yoghurt on a plate, load it up with the squash, scatter the nuts on top and enjoy with a green salad.

LEFTOVERS?
USE ANY LEFTOVER CHUNKS OF SQUASH IN THE ANY-VEG RISOTTO ON PAGE 113 OR IN THE FALAFELS ON PAGE 102 INSTEAD OF SWEET POTATO. YOU CAN ALSO KEEP THEM IN THE FRIDGE FOR UP TO FIVE DAYS, OR MASH THEM AND STORE IN THE FREEZER FOR THREE MONTHS.

KEEP THE SEEDS
FROM THE SQUASH –
SEE PAGE 186.

USE-IT-UP FRIED RICE.

PREP TIME: 10 MINUTES
COOK TIME: 30 MINUTES
SERVES: 4

350g basmati rice
½ white cabbage (500g)
6 spring onions (100g)
30ml sesame oil
200g frozen peas
150g peanuts
handful of fresh coriander
salt and pepper

FOR THE SAUCE
50ml sesame oil
40g honey or maple syrup
40ml soy sauce
thumb-size piece of fresh
   ginger, grated (50g)
2 garlic cloves, grated
1 tsp chilli flakes

SWAPS

FLAVOUR
• No cabbage? Use shredded
kale, whole mushrooms, small
florets of broccoli or cauliflower,
or sliced Brussels sprouts,
carrots, courgettes, fennel or
pepper.
• No frozen peas? Use
sweetcorn, edamame beans.

Ready to rummage? This one goes out to all the nearly-forgotten spring onions at the back of the fridge. Like the best kind of party, this crowd-pleasing dish makes any veg feel welcome. The sauce is super simple to make and is packed with sticky-sweet flavour. And, as cold cooked rice fries better than hot rice, you can use up any leftovers you already have. Wild.

1. Rinse the rice under cold water until the water runs clear. Then measure two parts water to one part rice into a pan - so for 300g of rice add 600ml of cold water.
2. Add a good pinch of salt, pop the lid on and bring to the boil. As soon as it's boiling, reduce to a simmer and cook for 8-10 minutes.
3. Turn off the heat, leave the lid on and allow the rice to steam for 5 minutes.
4. Meanwhile, slice the cabbage and cut each spring onion into 3-4 pieces. Throw them into a big saucepan with the sesame oil and fry over medium heat for 3-5 minutes.
5. To make the sauce, mix the sesame oil, honey, soy sauce, grated ginger, garlic and chilli flakes in a bowl.
6. Crank up the heat under the saucepan and add the cooked rice, frozen peas and sauce. Fry for 2-3 minutes. Don't worry if it catches a bit on the bottom of the pan - little crispy bits are good.
7. Roughly chop the peanuts and coriander and mix through the rice, saving a little for a garnish on top.

LEFTOVERS?

IF YOU'VE USED COLD, LEFTOVER RICE DON'T STORE IT AGAIN. BUT IF YOU'VE COOKED EVERYTHING AT THE SAME TIME, YOU CAN KEEP LEFTOVERS IN THE FRIDGE FOR UP TO THREE DAYS OR IN THE FREEZER FOR THREE MONTHS.

MARTYN'S CHUCK-IT-ALL-IN TAGINE.

PREP TIME: 20 MINUTES
COOK TIME: 35 MINUTES
SERVES: 4

50ml olive oil
2 white onions, chopped (300g)
thumb-size piece of ginger,
  finely chopped (30g)
3 garlic cloves, sliced
2 tbsp harissa paste (80g)
10 dried apricots, chopped
  (90g)
8 dates, pitted (80g)
2 tins tomatoes (800g)
300ml stock
1 tsp ground turmeric
2 tsp ground cinnamon
2 tins chickpeas, drained
  (450g)
4 large carrots, cut into chunks
  or cubes (500g)
300g green beans
handful of fresh coriander,
  chopped
salt and pepper

A tagine is a North African dish, named after the earthenware pot it's cooked in. But even if you don't have a traditional pot (we did tell you to clear out your cupboards), you can create a similar vibe with a big pan or slow cooker. Martyn's tagine uses classic flavours, but with the option to add pretty much any other veg you fancy. Serve with a flatbread (see page 76) or a steamy bowl of couscous.

1. Pour a good glug of oil into a saucepan and gently fry the onion and ginger for 3-5 minutes on a low heat.
2. Add the garlic, harissa paste, chopped apricots and dates and fry for 2-3 minutes.
3. Pour in the tinned tomatoes, stock, spices, chickpeas and carrots.
4. Bring to the boil then season with salt and pepper and simmer for 15-20 minutes.
5. Taste and add more spice or seasoning if needed.
6. To finish, slide in the green beans (whole) and simmer for a couple of minutes until soft.
7. Mix through the fresh coriander and get stuck in.

SWAPS

FLAVOUR
• No onions? Use leeks, fennel, peppers, aubergine.
• No carrots? Use potatoes, sweet potatoes, parsnips, squash, turnips.
• No green beans? Use spinach, broccoli florets.

LEFTOVERS?

KEEP IN THE FRIDGE FOR UP TO THREE DAYS AND IN THE FREEZER FOR THREE MONTHS.

SPICY-ROOT ROLL.

PREP TIME: 20 MINUTES
COOK TIME: 55 MINUTES
SERVES: 2

4 carrots (280g)
1 medium sweet potato (160g)
20ml olive oil
1 tsp curry powder
200g spinach
20g any butter
3 large filo sheets
100g goat's cheese, or feta,
   brie, Cheddar, vegan
   alternative
lemon zest (optional)
salt and pepper

SWAPS

FLAVOUR
• No carrots? Use beetroot,
peppers, aubergine, parsnips.
• No sweet potato? Use
potatoes, peeled squash or
pumpkin.
• No spinach? Use kale,
shredded cabbage, broccoli
florets.

MAKE LITTLE PARCELS
WITH THE REST OF THE
FILO SHEETS OR USE THE
FILLING IN THE SMOKY
VEG DIP ON PAGE 146.

LEFTOVERS?
KEEP THE FILLING IN
THE FRIDGE FOR THREE
DAYS OR IN THE FREEZER
FOR THREE MONTHS.

Roll up, roll up! This spicy veg roll uses ready-made filo pastry - a great staple to have in the fridge. Mix up the filling by adding anything you fancy, and don't be afraid to experiment with spices. We've used curry powder, but Martyn recommends trying Middle Eastern spices such as za'atar, or even just lemon and garlic. A simple salad is all you need on the side.

1. Preheat the oven to 180°C/160°C fan/350°F/gas 4.
2. Chop the carrots and sweet potato into large pieces roughly the same size. (No need to peel.)
3. Scatter both on a roasting tray. Then drizzle with oil and curry powder, season with salt and pepper and give everything a good mix.
4. Slide the roasting tray into the oven and roast for 20-30 minutes - give the tray a shake halfway through so the pieces roast evenly. You're aiming for nicely caramelised vegetables.
5. Meanwhile, cook the spinach in a pan of boiling water until soft. Allow to cool and squeeze out as much water as possible. Roughly chop into small pieces.
6. Melt the butter. Then lay a sheet of filo on a clean tea towel, brush with butter and lay another sheet on top. And repeat.
7. When the carrots and sweet potato are roasted, roughly chop them so they'll fit neatly on the pastry. Then lay them on top of the pastry along the shorter edge, leaving space at the sides. Top with the spinach and crumble over the cheese.
8. Season with salt, pepper and lemon zest, if you fancy it.
9. Roll up the pastry to form a log. Tuck the edges under to seal in the filling before the final roll.
10. Brush with any remaining butter, then lay the log on baking paper on a baking tray, seam side down, and bake in the oven for 20-25 minutes until golden brown.

THAI-INSPIRED RED CURRY.

PREP TIME: 10 MINUTES
COOK TIME: 35 MINUTES
SERVES: 4

2 aubergines (500g)
2 sweet potatoes (400g)
3 red peppers (640g)
500g firm tofu, or tempeh, seitan
50ml soy sauce
50ml sesame oil
100g red Thai curry paste
2 tins coconut milk (800ml)
150g rice vermicelli noodles, or rice, udon or soba noodles
1 lime
fresh coriander, to serve

SWAPS

FLAVOUR
• No aubergines, sweet potatoes, red peppers? Use cauliflower, courgettes, squash, swede, potatoes, celeriac, broccoli, Brussels sprouts.

Curry in a hurry? This mouth-wateringly simple dish uses a Thai curry paste - a really useful one to have in the cupboard. With a delicious base on the go, you're free to add almost any veg you have kicking about, and any substitute for the tofu, too. If you want to freeze a batch of the sauce, leave out the noodles and add them when you come to reheat.

1. Preheat the oven to 200°C/180°C fan/400°F/gas 6.
2. Chop your veg and tofu into similar-sized chunks, place on a large baking tray and drizzle over the soy sauce and sesame oil.
3. Give everything a really good mix and roast for 20-30 minutes, turning every now and again until golden brown.
4. Meanwhile, put the curry paste in a large saucepan and gently fry for 1-2 minutes to release the oils. Then add the coconut milk and gently simmer.
5. When your veg is ready, slide it into the sauce and gently mix together.
6. Drop the noodles into a bowl of room-temperature water and let them soak for 3 minutes.
7. Carefully pull them apart, then stir them into the curry until warm.
8. Divide the curry into bowls and top with freshly squeezed lime juice and roughly chopped coriander.

LEFTOVERS?
KEEP IN THE FRIDGE FOR
UP TO THREE DAYS AND
IN THE FREEZER FOR
THREE MONTHS.

ULTIMATE MAC & CHEESE.

PREP TIME: 30 MINUTES
COOK TIME: 35 MINUTES
SERVES: 4

200g macaroni (or any pasta)
1 head broccoli (400g)
150g frozen peas

FOR THE BECHAMEL
70g butter or vegan butter
50g plain flour
3 garlic cloves, thinly sliced
500ml any milk
150g Cheddar, grated, or vegan
  alternative
2 tsp Dijon mustard
pinch of nutmeg (optional)
2 tsp smoked paprika
crispy onions, for topping
  (optional)
salt and pepper

SWAPS

FLAVOUR
• No broccoli? Use green beans,
cauliflower, halved Brussels
sprouts.
• No peas? Use thinly-sliced
asparagus, grated courgettes,
tinned sweetcorn.

With mustard, paprika and crispy onions, this mac & cheese really packs a punch. It's also full of good-for-you green veg, and you're free to add more - roasted chunks of butternut squash, sweet potato, celeriac or parsnips - or anything else you've puréed. Want to prep it in advance? Blanch and refresh (see page 27) the macaroni and veg to reheat when you're ready.

1. Preheat the oven to 200°C/180°C fan/400°F/gas 6. Put a big pan of salted water to boil on the hob.
2. When the water boils drop in the macaroni and set a timer for 6 minutes.
3. Prepare the broccoli by cutting it into small florets and chopping the stem into thin discs.
4. After 6 minutes, add the broccoli and peas to the pan and boil for 2 more minutes, then drain.
5. To make the bechamel sauce, melt the butter in a saucepan over a low heat. (There's a little more butter than normal to help crisp up the garlic.)
6. Mix in the flour and garlic and cook for 5-8 minutes, until the garlic starts to crisp and turn golden.
7. Turn down the heat and slowly whisk in the milk until the sauce is nice and thick.
8. Mix through the cheese (saving a little for the top), mustard, nutmeg, if using, and paprika. Season to taste with salt and pepper.
9. Pour the bechamel sauce over the pasta, broccoli and peas. Give everything a good mix and slide into an ovenproof dish.
10. Pop in the oven for 15-20 minutes with a sprinkling of cheese on top, and bake until it's lightly golden and the cheese is bubbly. Add the crispy onions and enjoy.

GO AGAINST YOUR INSTINCTS
AND UNDERCOOK THE PASTA AND
BROCCOLI – IT ALL GETS BAKED, TOO.

LEFTOVERS?
KEEP IN THE FRIDGE FOR UP TO FIVE DAYS. IF YOU'RE FEELING SUPER FANCY, DON'T BAKE IT – LET IT SET IN THE FRIDGE THEN CUT INTO SQUARES, DIP INTO EGG, FLOUR AND BREADCRUMBS AND DEEP-FRY. WOW.

SIZZLING CELERIAC WITH HARISSA YOGHURT.

PREP TIME: 15 MINUTES
COOK TIME: 2 HOURS
SERVES: 2

zest and juice of 1 lemon
1 tsp ground turmeric
2 tsp smoked paprika
10g sugar
1/2 tsp ground cinnamon
40ml olive oil
1 celeriac (1kg)

FOR THE PICKLE
100ml red wine vinegar
50ml water
30g sugar
1 red onion, finely sliced (120g)
salt and pepper

FOR THE YOGHURT
1 tsp harissa paste
100g any yoghurt
60g flaked almonds, or walnuts,
   pine nuts, pecans

SWAPS

STRUCTURE
• No celeriac? Use cauliflower,
carrots, broccoli, parsnips,
leeks, sweet potato wedges
or halved cabbages (adjust
the cooking time for each
ingredient).

What could be simpler than whacking a whole vegetable in the oven to roast? Smothered in spices, this dish turns the humble celeriac (or whichever veg you're using) into a thing of wonder. And there's no need to eat it all in one go - keep any leftover chunks for soups, salads and curries. A week's meals in one.

1. Preheat the oven to 180°C/160°C fan/350°F/gas 4.
2. In a large mixing bowl, combine the lemon zest and juice, turmeric, paprika, sugar, cinnamon and oil, to form a wet paste. Give everything a really good mix and add a twist of salt and pepper.
3. Carefully stab the celeriac all over with a knife.
4. Put the celeriac in the mixing bowl and rub the paste all over. Take out the celeriac, drizzle more oil into the bowl and scrape any paste from the sides. Keep it for later.
5. Lay the celeriac on a baking tray, and cover it with foil so it will steam without burning. Roast in the oven for about an hour and a half, or until soft all the way through.
6. Meanwhile, make the pickle. Warm the vinegar, water and sugar in a saucepan until the sugar dissolves. Put the onion slices in a jar. Add a good pinch of salt to the onions and pour over the vinegar. Leave for 20 minutes while the celeriac is cooking.
7. Remove the foil from the celeriac and spread the leftover spice paste over it. Crank the oven up to 240°C/220°C fan/475°F/mark 9 and slide the celeriac back in for 5-10 minutes without the foil until it's nice and crispy.
8. Ripple harissa through the yoghurt and toast the almonds gently in a pan until golden brown.
9. Chop the celeriac into wedges. Serve with drizzled yoghurt, a sprinkle of toasted almonds and the pickled onions.

LEFTOVERS?

THE ROASTED CELERIAC WILL KEEP FOR THREE DAYS IN THE FRIDGE. YOU CAN ALSO MASH ANY LEFTOVERS TO SERVE AS A SIDE DISH, OR STIR THROUGH SALADS, SOUPS OR CURRIES. THE YOGHURT WILL KEEP IN THE FRIDGE FOR FIVE DAYS.

SNACKS AND SIDES.

LIMELIGHT-STEALING DISHES FOR PLATTERS OR PLUS-ONES.

RAW ZINGY SLAW.

PREP TIME: 20 MINUTES
SERVES: 4

½ red cabbage (400g)
1 red onion (150g)
2 carrots (150g)
2 apples (200g)
zest and juice of 2 limes
60ml olive oil
1 tsp chilli flakes
big handful of fresh coriander,
  chopped
salt and pepper

SWAPS

FLAVOUR
• No red cabbage? Use any
cabbage, kale, Brussels sprouts,
lettuce.
• No red onion? Use white
onion, fennel, cucumber.
• No apples? Use pears,
mangoes, pineapple, oranges.
• No carrots? Use courgettes,
beetroot.
• No coriander? Use any soft
herb - tarragon, basil, parsley,
chives.

LEFTOVERS?

THE SLAW WILL STAY
CRUNCHY FOR 1–2 DAYS IN
THE FRIDGE – YOU CAN
KEEP IT FOR UP TO FIVE
DAYS, BUT IT WILL START
TO GO SOFT.

Everyone needs a good slaw recipe in their lives, and this is one of the best. Light, fresh and with plenty of tang from the limes, plus a sweet kick from the apples, it makes a great addition to almost any dish. Use a mandolin, peeler or grater to chop everything into as fine or as chunky pieces as you like – it's up to you.

1. Chop the cabbage, peel and slice the onion, grate the carrots, remove the core from the apples, then thinly slice them.
2. Put everything in a bowl and add the lime zest and juice. Then add the olive oil, chilli flakes and a big handful of chopped coriander. Give everything a really good scrunch, and season with salt and more lime if needed.
3. Serve the slaw straight away so it keeps its crunch.

TRY PICKLING THE
OTHER HALF OF THE RED
CABBAGE – SEE PAGE 30.

LEFTOVER RISOTTO ARANCINI.

PREP TIME: 20 MINUTES
COOK TIME: 10 MINUTES
SERVES: 2

400g leftover risotto - see page 113
100g mozzarella cheese, or any soft cheese
100g plain flour
2 eggs, lightly beaten
100g breadcrumbs
vegetable oil, for deep-frying

LEFTOVERS?

KEEP YOUR ARANCINI IN THE FRIDGE FOR UP TO THREE DAYS AND IN THE FREEZER FOR THREE MONTHS. TO REHEAT, POP INTO THE OVEN AND MAKE SURE THEY'RE PIPING HOT ALL THE WAY THROUGH BEFORE SERVING.

Arancini are deep-fried rice balls originating from Sicily. Ours are made from leftover risotto - see page 113 - and are stuffed with mozzarella to make them even more gooey. You can also make fresh risotto to use here - just allow it to cool down before you squish it into balls. And if you don't have any mozzarella, try stuffing the balls with a different cheese, or even add sun-dried tomatoes or ragu.

1. Roll the rice into balls roughly the size of golf balls. Break off small pieces of mozzarella and push them into the middle of the balls, reshaping them to cover the cheese.
2. Prepare three bowls with flour, beaten egg and breadcrumbs.
3. Gently roll the balls in the flour, dusting off any excess, then dip them first into the eggs then the breadcrumbs. Repeat the egg and breadcrumb stages if you have enough of them.
4. Heat the oil in a deep saucepan or wok. If you have a thermometer, you're aiming for 180°C/350°F. If not, sprinkle in some breadcrumbs to see if they bubble and fizz (be very careful).
5. When the oil is hot enough, carefully drop in the balls and deep-fry for 3-5 minutes until golden all over. Or you can bake them in the oven at 180°C/160°C fan/350°F/gas 4 for 10-12 minutes until golden brown.
6. Serve the arancini with garlic mayo, if you have some, spicy tomato chutney or a small bowl of fresh pickles.

FOR SPEEDY BREADCRUMBS, MAKE A FEW PIECES OF TOAST, ALLOW THEM TO COOL, THEN BLITZ IN THE FOOD PROCESSOR.

SALAD ROLLS.

PREP TIME: 20 MINUTES
COOK TIME: 5 MINUTES
SERVES: 4 (MAKES 4-6)

1 carrot (75g)
¼ cucumber (100g)
60g lettuce
100g firm tofu
50g rice vermicelli noodles
30g peanut butter
20ml soy sauce
20ml vinegar
1 garlic clove, grated
½ tsp chilli flakes
8 sheets of rice paper (16cm
 diameter paper)
small handful of fresh mint

SWAPS

FLAVOUR
• No carrot or cucumber? Use
beetroot, fennel, courgettes,
peppers.
• No lettuce? Use white
cabbage, red cabbage, avocado,
radishes.

Inspired by Vietnamese summer rolls, these super-fresh treats are a great way to use up any salad or veg you have in the fridge drawer. And if you have any leftover ingredients at the end, you can turn them into a salad or stir-fry with the dipping sauce as a dressing. Yep, that kind of wild, waste-free behaviour is just the way we roll.

1. Prep the veg and tofu by chopping everything finely. You can do this however you like: grate, shred, cut into ribbons or thin strips or even into julienne (a fancy word for thin matchsticks).
2. Follow the instructions on the noodle packet – most say to put the noodles in a heatproof bowl, cover with boiling water and gently move around. After 5 minutes, drain and set to one side.
3. Make the dipping sauce by mixing the peanut butter, soy sauce, vinegar, grated garlic and chilli flakes in a bowl. Add a little hot water to loosen the sauce if it looks too thick.
4. Soak the rice paper sheets in warm water for 10-20 seconds until soft, carefully pull out of the water and lay on a clean work surface.
5. Fill the rice paper sheets with the veg and tofu – don't overfill or they'll be hard to fold. Fold the paper over the filling, and pull it in tightly. Then fold in the sides and roll up. Keep making rolls until you've used all the filling.
6. Serve with the dipping sauce on the side.

LEFTOVERS?
THE CHOPPED VEG WILL KEEP
IN THE FRIDGE FOR UP TO A
WEEK AND THE NOODLES WILL
KEEP IN THE FRIDGE FOR UP TO
THREE DAYS.

EASY-ISH HUMMUS.

PREP TIME: 5 MINUTES
COOK TIME: 30 MINUTES
SERVES: 2

1 red pepper
olive oil, for drizzling
1 tin chickpeas, drained (240g)
1 tsp ground cumin
80g tahini
1 garlic clove, grated
½ tsp lemon juice
salt and pepper

TO SERVE
mixed sliced raw veg for
  crudités (cucumber, celery,
  courgette, radishes)
Cracking crisps - page 182

SWAPS

FLAVOUR
• No pepper? Use sun-dried
tomatoes, or roasted beetroot
or carrots.

Why easy-ish? Well, Martyn has given two options for making this heavenly hummus - the first is slightly more labour intensive and is inspired by Ottolenghi's approach. The second, which leaves out a step, is faster if you need to whip up a dip in a hurry. Take your pick. For a silky-smooth texture, follow every step. If you're in a hurry, skip step 3.

1. Preheat the oven to 200°C/180°C fan/400°F/gas 6.
2. Thinly slice the pepper, drizzle with a little oil, salt and pepper and roast for 20-30 minutes.
3. Pour the chickpeas onto a clean tea towel and gently rub to remove the skins. Be gentle - you're not trying to squash them.
4. Put the chickpeas in a saucepan, cover with boiling water, add a pinch of salt then simmer for 10-12 minutes until soft. Drain the chickpeas and keep the water.
5. Put the chickpeas into a blender or food processor with the cumin, tahini, grated garlic and lemon juice.
6. Start blitzing, scraping down the sides as you go. Add a tablespoon of the chickpea water, then taste and add more tahini and lemon juice if needed.
7. Add the pepper strips and blitz for a couple of minutes until really smooth.
8. Scoop the hummus into a bowl, drizzle with a little olive oil and serve with crisps or veg crudités.

HUMMUS IS A GREAT
ALTERNATIVE TO FETA
IF YOU'RE MAKING THE
CHARGRILLED VEG RECIPE
ON PAGE 148.

CHARGRILLED VEG WITH WHIPPED FETA.

PREP TIME: 10 MINUTES
COOK TIME: 30 MINUTES
SERVES: 2

1 courgette (160g)
1 aubergine (320g)
40ml olive oil
100g green olives, pitted
1 garlic clove, grated
juice and zest of 1 lemon
small bunch of fresh parsley,
   chopped
200g feta or vegan alternative
1-2 tbsp cream cheese or vegan
   alternative
salt and pepper (optional)

SWAPS

FLAVOUR
• No courgettes or aubergines?
Use broccoli or cauliflower
florets, fennel wedges or big
chunks of pepper.

Martyn says: 'This recipe uses chargrilled olives - a dish I learnt from an Italian friend. They're outrageously good at helping to build flavour, as the saltiness takes a back seat and the sweetness comes through. You can chargrill a huge variety of veggies for this dish - get some nice colour using a chargrill pan, then whack everything in the oven if needed. Incredible.'

1. Slice the courgette and aubergine lengthways into thin strips, drizzle with a little oil and lay the strips on the chargrill pan. Cook for 1-2 minutes on both sides then pop into a bowl.
2. Take the olives and press them down on a chargrill pan with a spatula until you have char marks on both sides. Then chop them roughly and put in the bowl with the veg.
3. Add the olive oil, grated garlic, lemon zest and juice and parsley and give everything a good mix.
4. In a blender, blitz the feta and cream cheese together. You're after a super-smooth texture so add more cream cheese if needed. Have a taste and add salt and pepper.
5. Load up a plate with a big smear of the whipped feta and pile the chargrilled veg on top.

LEFTOVERS?
THE MARINATED VEGETABLES WILL KEEP
FOR UP TO FIVE DAYS IN THE FRIDGE. THEY
CAN BE USED IN THE ROASTED-VEG BAGELS
(PAGE 59) OR AS A TOPPER FOR THE LOADED
FLATBREADS WITH HUMMUS (PAGE 76).

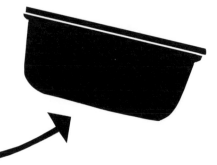

MASH CROQUETTES.

PREP TIME: 35 MINUTES
COOK TIME: 20 MINUTES
SERVES: 2 (MAKES ABOUT 9)

FOR THE MASH
2 floury potatoes (350g)
80g Cheddar cheese, grated
80g mozzarella, grated
1 heaped tsp mustard (20g)
salt and pepper

FOR THE COATING
80g plain flour
2 eggs, lightly beaten
80g breadcrumbs
vegetable oil, for deep-frying

SWAPS

STRUCTURE
• No potatoes? Use sweet potatoes, celeriac, parsnips, peeled pumpkin or butternut squash.

LEFTOVERS?

CROQUETTES ARE PERFECT FOR THE FREEZER. IF YOU'RE COOKING A BIG BATCH, MAKE THE MIXTURE, SHAPE AND COAT THE CROQUETTES, SEPARATE THEM ON A TRAY AND POP IN THE FREEZER. AFTER A FEW HOURS WHEN THEY'RE SET HARD, PUT THEM INTO AN AIRTIGHT CONTAINER AND FREEZE FOR THREE MONTHS.

Made too much mash? It happens. In fact, potatoes are one of the most thrown-away ingredients – around 4.4 million are wasted in the UK every day. These super-easy croquettes are great for using up whatever's left in the pan, or you can whip up a fresh batch with any mashable veg in the fridge.

1. Start by making plain mash. Cut the potatoes, skin-on, into even cubes and pop them into a pan of salted water. Bring to the boil and simmer for about 8-10 minutes until the potatoes are turning soft.
2. Drain the potatoes, allow them to steam then mash them in a bowl.
3. Mix through the cheeses and mustard and season with salt and pepper.
4. Shape the mixture into cylinders or balls and put them in the fridge.
5. Put the flour, egg and breadcrumbs into three separate bowls.
6. Roll the croquettes in the flour, dip them in the egg then coat them in breadcrumbs. If you have enough egg and breadcrumbs, repeat the stages.
7. Heat a saucepan of oil. If you have a thermometer, you're aiming for 180°C/350°F. If not, sprinkle in some breadcrumbs to see if they bubble and fizz (be very careful).
8. When the oil is hot enough, deep-fry the croquettes for 3-4 minutes or until golden brown. Or bake them in the oven at 200°C/180°C fan/400°F/gas 6 for 8-10 minutes.
9. Serve the croquettes with garlic mayo if you have any, or tomato chutney.

MIXED-VEG TEMPURA.

PREP TIME: 10 MINUTES
COOK TIME: 5 MINUTES
SERVES: 2-4

2 small sweet potatoes (240g)
200g broccoli florets
1 red pepper (300g)
vegetable oil, for deep-frying

FOR THE BATTER
1 egg (or see vegan batter
   recipe below)
200ml ice-cold water (or
   sparkling water)
140g cornflour
60g plain flour
pinch of salt

FOR THE VEGAN BATTER
150g self-raising flour
pinch of salt
ice-cold water

FOR THE DIPPING SAUCE
60ml soy sauce
30ml sesame oil
juice of 1 lime
20g sugar

SWAPS

FLAVOUR
• No peppers? Use asparagus,
green beans, kale, halved
Brussels sprouts.
• No broccoli? Use cauliflower.
• No sweet potatoes? Use
carrots, peeled butternut
squash or pumpkin.

Tempura is a Japanese dish with lightly battered and deep-fried ingredients. Not sure you've got what it takes? Our version is nice and simple, and with these top tips you can't go wrong. One: keep the batter cold using iced water. Two: don't overmix – a few lumps are fine. Three: use it all straight away. Four: cut the veg into thin, similar-sized bits. See? Now get stuck in.

1. Preheat the oven to its lowest temperature.
2. Prep the veg. Cut the sweet potatoes into $\frac{1}{2}$ cm slices, the broccoli into mini florets and the pepper into bite-sized pieces.
3. Heat a saucepan of oil while you prepare the batter. You need enough oil to let the veg swim happily, so make sure the pan is big enough to prevent the oil spilling over.
4. To make the batter, mix the egg and ice-cold water in a bowl, then add both flours and a pinch of salt and give it a very quick mix. (Don't overdo it.) To make vegan batter, mix the flour and a pinch of salt with a little ice-cold water at a time until you have a smooth and thick batter.
5. Test the temperature of the oil. If you have a thermometer, you're aiming for 180°C/350°F. If not, drop in a bit of veg to see if it bubbles and fizzes. (Be very careful.)
6. Dip the vegetable pieces into the batter one by one and then carefully drop them into the oil, moving quickly from batter to pan. Make sure the pan isn't overcrowded.
7. Cook for 45-60 seconds or until the vegetables are floating. Then roll them over and cook for a further 30 seconds.
8. Drain the vegetables on a paper towel and lay them in the oven to keep warm while you fry the rest.
9. For the dipping sauce, mix the soy sauce, sesame oil, lime juice and sugar.
10. Serve the tempura veg with the dipping sauce in a little bowl.

LEFTOVERS?

TEMPURA IS BEST SERVED FRESH, BUT YOU CAN FREEZE THE DEEP-FRIED VEG (ONCE COOL) FOR THREE MONTHS. THEY CAN BE REHEATED IN THE OVEN AT 180°C/160°C FAN/350°F/GAS 4 FOR 8-10 MINUTES OR UNTIL HOT THROUGH. THE OIL CAN BE REUSED ABOUT EIGHT TIMES – ALLOW IT TO COOL, STRAIN OUT ANY BITS AND STORE IN AN AIRTIGHT CONTAINER.

LEFTOVERS?
THE BURNT BUTTER CAN BE STORED IN A CONTAINER IN THE FRIDGE FOR WEEKS – ADD IT TO DISHES THAT INCLUDE BUTTER IN THE RECIPE TO GIVE THEM AN EXTRA NUTTY HIT.

BURNT-BUTTER SPUDS.

PREP TIME: 5 MINUTES
COOK TIME: 20 MINUTES
SERVES: 2

60g butter
400g new potatoes
50g pine nuts
olive oil, for frying
big bunch of fresh basil,
  chopped
salt and pepper

SWAPS

FLAVOUR
• No new potatoes? Use any
green veg, carrots, cauliflower
florets, Brussels sprouts
(adjust the blanching time).

This simple side dish is proof that even the most everyday vegetables can be utterly transformed with just a few ingredients. Burnt butter with sage is a classic dish - if you want to give that a go, stir sage into the butter while it's still warm. Here, Martyn suggests letting the butter cool before mixing with fresh basil to create 'something truly epic'.

1. Gently melt the butter in a large saucepan, swirling it round every now and again. Keep your eye on it as it starts to foam and smell nutty - you're aiming for a light brown colour, so turn the heat down if it's burning too fast.
2. When the butter is light brown, remove the pan from the heat and pour the butter into a bowl so it stops cooking.
3. Meanwhile, slice the potatoes and blanch in boiling water for 8-10 minutes until just turning soft, then drain.
4. Gently toast the pine nuts in a dry frying pan or in the oven at 180°C/160°C fan/350°F/gas 4 for 5 minutes, until golden brown. Watch they don't burn.
5. Pop the potatoes into a frying pan with a little oil and sauté for a few minutes until they start to go crispy.
6. To finish, add the burnt butter, pine nuts and fresh basil to the potatoes with a good twist of salt and pepper.

SMOKY VEG DIP.

PREP TIME: 5 MINUTES
COOK TIME: 40 MINUTES
SERVES: 2

4 carrots (300g)
20ml olive oil, plus extra for
   drizzling
1 1/2 tsp smoked paprika
15ml honey or maple syrup
handful of fresh parsley
20g tahini
10ml white wine vinegar
1 tsp caraway seeds (optional)
   mixed raw veg for crudités
   (cucumber, celery, courgettes,
   radishes) or crispbread
salt and pepper

SWAPS

FLAVOUR
• No carrots? Use parsnips,
cauliflower, celeriac, aubergine,
courgettes, beetroot, peppers.

Smooth or chunky? Spicy or mild? Carrots or parsnips? There are so many ways of making this dip your own, and so many other dishes that can help with the ingredients. Throw in celeriac from page 126. Roasted veg from the breakfast bagels on page 59. Or a few chunks of anything lurking in the freezer. Then dip to your heart's content.

1. Preheat the oven to 180°C/160°C fan/350°F/gas 4.
2. Cut the carrots into rough chunks and lay on a roasting tray. Drizzle with a little oil, salt and 1 teaspoon of the paprika. Slide the tray into the oven and cook for 20-30 minutes or until just starting to look roasted.
3. Drizzle the carrots with honey and give them a good mix, then pop back into the oven and crank up the heat to 220°C/200°C fan/425°F/gas 7 for 5-10 minutes. You're aiming for slightly burnt, caramelised carrots.
4. In a bowl, whisk together the oil, parsley, tahini, the remaining paprika and vinegar with a good pinch of salt.
5. Toast the caraway seeds in a dry pan for 1-2 minutes or until they start to pop. Then tip into the tahini mixture and stir through.
6. Blitz or roughly chop the roasted carrots, then add the tahini mixture. Season to taste and enjoy with crudités or crispbread.

> ### LEFTOVERS?
> IF YOU BLITZED THE DIP, TRY MIXING IT THROUGH A SOUP TO ADD A SMOKY FLAVOUR. IF YOU KEPT IT CHUNKY, USE IT AS A SALAD DRESSING. ANY EXTRA CAN BE KEPT IN THE FRIDGE FOR UP TO FIVE DAYS OR FROZEN FOR THREE MONTHS.

TO DIAL UP THE FLAVOUR, ADD SMOKED SALT AND CHILLI.

CURRIED VEG WITH YOGHURT & PICKLES.

PREP TIME: 5 MINUTES
COOK TIME: 20 MINUTES
SERVES: 2

2 carrots (180g)
20ml olive oil
300g green beans, whole
50g any butter
2 tsp curry powder
½ tsp coriander seeds
1 garlic clove, thinly sliced
80g any yoghurt
mixed pickles (fennel, onion,
  cauliflower stalk)

SWAPS

FLAVOUR
• No carrots? Use parsnips,
cauliflower, celeriac.
• No green beans? Use any
green veg, peas.

Watch out, main courses – this tangy side dish is set to steal the limelight. Beautifully balanced with warm spices, creamy yoghurt and sharp pickles, the flavours far outweigh the effort of making it. If you haven't made a pickle yet, have a look at page 30 to create your topping.

1. Chop the carrots into bite-sized pieces and blanch in boiling water for 1–2 minutes, until they soften slightly. Then drain and pop straight into a frying pan with a little oil.
2. Sauté the carrots for a couple of minutes, turning them every now and then. Be patient and give them time to catch a little to bring out the colour and flavour.
3. When the carrots are looking good, drop your beans into the pan with the butter, curry powder, coriander seeds and garlic.
4. Gently fry over medium heat until the butter goes frothy, but don't let it burn.
5. Dollop a big spoonful of yoghurt on to a plate and make a little well to catch the butter, then spoon in the veg and butter and top with fresh pickles.

MADE THE SAUTEÉD SPUDS ON
PAGE 145? YOU CAN USE ANY
LEFTOVER BURNT BUTTER IN
THIS RECIPE TOO.

LEFTOVERS?

IF YOU MAKE A BIG BATCH
FOR A DINNER PARTY, SERVE
THE YOGHURT AND VEG
SEPARATELY SO YOU CAN USE
ANY LEFTOVER VEG IN A SOUP
OR CURRY. THEY'LL KEEP IN
THE FRIDGE FOR UP TO THREE
DAYS OR IN THE FREEZER FOR
THREE MONTHS.

TANGY MISO GREENS.

PREP TIME: 5 MINUTES
COOK TIME: 5 MINUTES
SERVES: 2

10g miso paste
20g honey or maple syrup
40ml sesame oil
20ml soy sauce
small piece of fresh ginger,
    grated (15g)
1 garlic clove, grated
¼ tsp chilli flakes, or to taste
2 pak choi (250g)
200g tenderstem broccoli

TO SERVE (OPTIONAL)
crispy garlic, crispy onions,
toasted sesame seeds

SWAPS

FLAVOUR
• No pak choi or broccoli? Use
shredded kale, cavolo nero or
cabbage, whole green beans
or asparagus, halved Brussels
sprouts, cauliflower florets,
peas, edamame beans.

Looking for a last-minute, cupboard-friendly side dish?
Gather your ingredients and get cracking. A word of warning:
once you've made the sauce, the greens cook so quickly that
you want to make sure everything else is dished up and ready
to go. Nobody wants limp pak choi.

1. Start by making the sauce. In a bowl, mix the miso paste,
honey, 30ml of the sesame oil, the soy sauce, grated ginger
and garlic and chilli flakes.
2. Slice the pak choi into quarters and leave the tenderstem
broccoli whole.
3. Heat a pan on a high heat then pour in the remaining 10ml
(a good glug) of sesame oil and carefully add your pak choi
and broccoli.
4. Fry over a high heat for 1-2 minutes, then add a splash of
water and pop on a lid. Steam for 1-2 minutes until the water
has evaporated.
5. Pour over the miso sauce and coat all the vegetables. (Try
not to overcook the veg - you want them to have a bit of
crunch.)
6. To finish, sprinkle on some crispy garlic, onions or
toasted sesame seeds, if you like. Serve as the perfect
accompaniment to the Thai-inspired red curry on page 123.

LEFTOVERS?
MAKE A BIG BATCH OF THE SAUCE – IT
WILL KEEP IN THE FRIDGE FOR A COUPLE
OF WEEKS AND IS A GREAT UMAMI HIT
TO ADD TO RICE, NOODLES AND SALADS.
THE VEG WILL KEEP IN THE FRIDGE FOR
UP TO THREE DAYS.

SUPER-QUICK PAKORAS.

PREP TIME: 15 MINUTES
COOK TIME: 10 MINUTES
SERVES: 2

vegetable oil, for frying
150g chickpea flour (also known
  as gram flour)
1½ tsp mild curry powder
100ml water
1 onion (150g)
1 medium potato (100g)
¼ cauliflower (100g)
salt and pepper

FOR THE SAUCE
handful of fresh mint
80g any yoghurt

SWAPS

FLAVOUR
• No onion? Use leek, fennel,
red onion.
STRUCTURE
• No potato? Use sweet potato,
peeled butternut squash,
celeriac, parsnips.
• No cauliflower? Use beetroot,
courgettes, broccoli.

LEFTOVERS?
THE BATTER WILL KEEP FOR
1-2 DAYS IN THE FRIDGE. ONCE
COOKED, THE PAKORAS WILL
KEEP FOR 2-3 DAYS IN THE
FRIDGE OR IN THE FREEZER
FOR THREE MONTHS. THEY
CAN BE REHEATED IN THE
OVEN UNTIL CRISPY.

Pakoras are Indian spiced fritters - a simple snack, starter or side. You can use almost any veg you fancy. And feel free to mix up the spices - the recipe uses mild curry powder, but turmeric, cumin, coriander and fenugreek would be delish too.

1. Preheat the oven to its lowest setting. You can deep-fry or shallow fry, so choose how much oil you want to use and make sure the pan is big enough to stop it spilling over. If you're deep-frying, heat a heavy-bottomed saucepan of oil on the hob.
2. In a bowl, mix the chickpea flour, curry powder and a pinch of salt.
3. Slowly mix the water into the flour to form a thick batter.
4. Grate the veg using a large box grater or shred with a knife. Include the stem and leaves of the cauliflower.
5. Mix the vegetables through the batter, so it sticks everything together.
6. Shape the pakoras in your hands into mini patties with about a tablespoon of batter in each. Keep them a little rough - little bits of veg poking out will give you nice crispy edges.
7. If you're deep-frying, test the temperature of the oil. If you have a thermometer, you're aiming for 180°C/350°F. If not, drop in a little bit of veg to see if it bubbles and fizzes. (Be very careful.) If you're shallow frying, heat a splash of oil in a frying pan.
8. Carefully drop the pakoras into the oil and fry for 2-3 minutes until golden brown, rolling them around to make sure they cook evenly.
9. Drain on a paper towel and pop in the oven to keep warm while you cook the rest.
10. For the dipping sauce, blitz the mint and yoghurt with a blender to a smooth consistency.
11. Serve the hot pakoras with the dip and enjoy.

DON'T TRY TO CHOP YOUR VEG INTO EVEN-SIZED CHUNKS – THE WILDER THE BETTER!

SWEET STUFF.

FOR DESSERTS, AFTERNOON TEAS OR TUESDAYS.

164

166

156

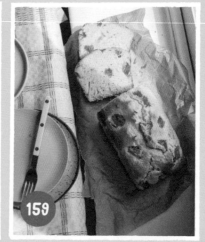

159

170

160

FRUITY TIRAMISU POTS.

PREP TIME: 15 MINUTES
COOK TIME: 15 MINUTES
REST TIME: 3 HOURS
SERVES: 4

FOR THE COMPOTE
500g cherries, pitted
1 tbsp honey

FOR THE TIRAMISU CREAM
2 eggs
60g caster sugar
200g mascarpone
1 tsp amaretto liqueur

FOR LAYERING
100g ladyfinger biscuits
1 tsp cocoa powder, to serve

SWAPS

FLAVOUR
• No cherries? Use
strawberries, raspberries,
plums, peaches.

Fruit compote. Squishy biscuits. And a light mascarpone with a sprinkling of cocoa powder. Nothing says 'I've made an effort' like a beautifully layered pud. (Especially when each layer tastes this dreamy.) We've used cherries, but you can use any fruit compote, or even just chopped fresh fruit.

1. Start by making the cherry compote. Put the cherries and honey in a small saucepan. Cook for 10-15 minutes over medium heat, until the fruit is soft and has released its juice. Turn off the heat and allow to cool.
2. Next, make the cream. Separate your eggs and place the yolks in one mixing bowl and the whites in another. Add the sugar to the bowl with the yolks, and whisk with an electric mixer for 2-3 minutes, until pale and fluffy.
3. Add the mascarpone and amaretto to the yolk mixture. Mix until combined.
4. Now whisk the egg whites into firm peaks. Fold the egg whites into the mascarpone mixture to form a light and fluffy cream.
5. Layer glasses or jam jars with the cherry compote, broken ladyfinger biscuits and mascarpone cream. Cover and chill in the fridge for three hours or overnight.
6. To serve, dust with the cocoa powder and dig in.

NOT SURE WHAT FOLDING
MEANS? USE A SPATULA
TO GENTLY LIFT THE TWO
ELEMENTS IN THE BOWL,
TURNING THEM OVER ON TOP
OF EACH OTHER UNTIL THEY'RE
COMPLETELY MIXED.

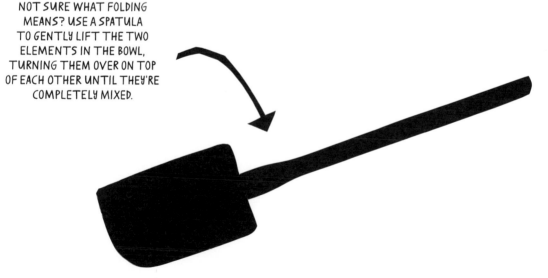

LEFTOVERS?
KEEP THE TIRAMISU IN
THE FRIDGE FOR UP TO
TWO DAYS.

CAMILLE'S YOGHURT CAKE.

PREP TIME: 15 MINUTES
COOK TIME: 50 MINUTES
SERVES: 6-8

1 pot yoghurt (125g); use the
   pot or a ramekin for
   measuring
3 pots self-raising flour
1 pot sugar
½ pot vegetable oil, plus extra
   for greasing
3 eggs
2 grapefruit

SWAPS

FLAVOUR
• No grapefruit? Use oranges,
clementines, mandarins,
strawberries, raspberries,
cherries, peaches, apricots,
bananas.

LEFTOVERS?
TRY MAKING YOGHURT CAKE
FRENCH TOAST – DIP SLICES OF
YOGHURT CAKE INTO A BATTER
MADE FROM EGG, SUGAR AND
MILK OF YOUR CHOICE. THEN
PAN-FRY IN BUTTER UNTIL
GOLDEN AND CRISP.

Camille says: 'Like most kids, I learnt to bake with my grandma. This fluffy yoghurt cake is the first one I made with her – now I see why! You use the yoghurt pot for measuring instead of scales, so it's the perfect recipe for children. I've used grapefruit but feel free to use any other fruit – I love it made with berries, peaches and plums. Just chop and fold the fruit directly into the batter.'

1. Preheat the oven to 180°C/160°C fan/350°F/gas 4.
2. Put the yoghurt, flour, sugar, oil and eggs into a large mixing bowl. (Use the yoghurt pot to measure ingredients, or a ramekin if your yoghurt came in a bigger pot.)
3. Mix to combine, then add the zest from the grapefruit.
4. Next peel the grapefruit and chop the flesh into small pieces. Add the flesh to the batter and mix to combine.
5. Pour the batter into a lightly greased 900g loaf tin and bake for 40-50 minutes, or until a knife inserted in the middle comes out clean.
6. Set aside to cool slightly then remove from the tin, slice and tuck in.

SUMMER PAVLOVA.

PREP TIME: 15 MINUTES
**COOK TIME: 1 HOUR
30 MINUTES**
SERVES: 4-6

FOR THE MERINGUE
5 egg whites, at room
  temperature
250g caster sugar
2 tsp cornflour
½ tsp vanilla extract, or almond
  extract, orange blossom water
  or rose water

FOR THE FILLING
250ml double cream
200g strawberries
icing sugar, for dusting

SWAPS

FLAVOUR
• No strawberries? Use
raspberries, blueberries,
blackberries, passionfruit,
mangoes, peaches, kiwis.

LEFTOVERS?

TO TURN THEM INTO
ETON MESS, BLITZ
THE FRUIT TO MAKE
A QUICK COULIS, THEN
LAYER IT IN GLASSES
OR JAM JARS WITH
BROKEN MERINGUE AND
WHIPPED CREAM.

Never made meringue before? Now's your chance. This recipe makes it super easy and, best of all, we think the wonkier it turns out, the better. Pile your creation high with whipped double cream and fresh chopped fruit of your choice. Then invite all your friends round to show off your new skills.

1. Preheat the oven to 140°C/120°C fan/275°F/gas 1, and line a large baking tray with baking paper.
2. Place the egg whites in a dry, clean, grease-free mixing bowl. This is important and will help you turn the egg whites into a lovely meringue.
3. Use an electric mixer to whisk the egg whites until firm peaks form. This can take 4-5 minutes.
4. Gradually add the sugar, a tablespoon at a time, whisking constantly, until the mixture is glossy and the sugar has dissolved. This can take 3-4 minutes.
5. Add the cornflour and vanilla extract. Whisk until just combined.
6. Spoon the meringue into the centre of the lined baking tray to form a circular mound. Use the back of the spoon to make a nest shape.
7. Bake for 1 hour 30 minutes, then turn off the heat, open the oven door and let the meringue cool completely.
8. When the meringue is cool and you're ready to serve, put the double cream into a mixing bowl and whisk until it thickens. Spoon it into the meringue nest and top with fresh strawberries and a dusting of icing sugar.

MELON JAM TART.

PREP TIME: 15 MINUTES
COOK TIME: 1 HOUR
20 MINUTES
SERVES: 4-6

1 melon (cantaloupe or galia)
80g golden caster sugar
1 vanilla pod
1 packet ready-rolled shortcrust
  or vegan pastry (320g)
handful of flaked almonds
  (optional)

FOR THE MELON SEEDS
melon seeds
pinch of ground cinnamon
1 tsp vegetable oil

SWAPS

FLAVOUR
• No melons? Use roughly 400g
of peaches, plums, strawberries.

LEFTOVERS?

SPREAD ANY
LEFTOVER JAM
ON TOAST FOR
BREAKFAST. KEEP
THE TART IN
AN AIRTIGHT
CONTAINER IN A COOL
PLACE FOR UP TO
THREE DAYS.

Melon jam? Whoever heard of such a thing? Trust us on this one – mixed with vanilla, the melon softens to create the sweetest, scrummiest jam in the land. If you fancy a more traditional tart, or if you have other fruit to use up, make the jam with peaches, plums or strawberries instead.

1. Peel, seed and chop the melon. Compost the peel but keep the seeds.
2. Place the melon chunks in a saucepan along with the sugar. Cut the vanilla pod lengthways and use a small knife to carefully scrape out the seeds. Add both seeds and pod to the pan.
3. Place the pan over a medium heat. Cook for about 45 minutes, stirring regularly, or until you have a liquid jam consistency.
4. In the meantime, preheat the oven to 180°C/160°C fan/350°F/gas 4. Place the pastry in a tart tin, then fill it with a circle of baking paper and add baking beans (or rice) to weigh it down.
5. Bake for 15 minutes, then carefully remove the paper and beans and cook for another 5 minutes. (This is called blind baking and helps to seal the surface and stop the pastry going soggy.)
6. Place the melon seeds in a saucepan, cover with water and boil for 3 minutes. Drain, pat dry and place in a baking dish. Add the ground cinnamon and oil. Toss to coat then bake for 4-5 minutes until golden. Remove from the oven and set aside to cool.
7. Pour and spread the melon jam over the pastry. Sprinkle over the flaked almonds and slide the tin back into the oven for 25-35 minutes, or until golden brown.
8. Serve with the melon seeds on top, and a big scoop of whatever you fancy.

ZESTY COOKIES.

PREP TIME: 1 HOUR
COOK TIME: 15 MINUTES
MAKES: 8-10 COOKIES

FOR THE COOKIES
zest of 1 lemon
50g honey
35g soft brown sugar
60g butter, at room
   temperature
pinch of salt
200g plain flour, plus extra for
   dusting
1 tsp baking powder
1 tsp ground aniseed
1 egg

FOR THE GLAZE
150g icing sugar
juice of 1 lemon

SWAPS

STRUCTURE
• No lemon? Use lime, orange,
clementine.

LEFTOVERS?
STORE THE COOKIES
IN AN AIRTIGHT
CONTAINER IN A COOL
PLACE FOR UP TO
THREE DAYS.

Glazed with a sweet layer of lemony icing sugar, these soft and crumbly cookies are the ultimate afternoon treat. (Or morning, or lunchtime, or midnight.) The recipe has a few more steps than other baked things in the book, but don't let that put you off – the results are totally worth it.

1. Put the lemon zest, honey, sugar, butter and salt in a large mixing bowl. Beat with an electric mixer until fluffy.
2. Stir in the flour, baking powder and ground aniseed. The mixture will have a crumbly consistency.
3. Add the egg and mix in well. Then use clean hands to mix until the dough comes together into a ball. Don't overmix or the dough will get stickier.
4. Place the dough between two sheets of baking paper and roll to 1.5cm thick. Lay on a baking tray and chill in the fridge for 40 minutes.
5. Preheat the oven to 200°C/180°C fan/400°F/gas 6, and take the chilled dough out of the fridge.
6. Remove the top layer of baking paper and cut out 8-10 cookies, depending on the size of your cookie cutter. Reroll the small leftover pieces of dough into another cookie.
7. Bake the cookies in the oven for 15-17 minutes, until slightly golden. Then remove and transfer to a cooling rack. Leave until they're room temperature.
8. In the meantime, make the glaze. In a small bowl, mix the icing sugar and half the lemon juice. Add extra lemon juice a little at a time and stop when the glaze is thick and opaque. (If the glaze looks a bit thin, add extra icing sugar.)
9. Place a tray under the cooling rack to catch the drips and dip each cookie into the glaze to add a layer of icing on top. Put them back on the cooling rack.
10. Leave the cookies to set at room temperature for at least an hour, so the icing hardens.

NO-BAKE TROPICAL CHEESECAKE.

PREP TIME: 20 MINUTES
REST TIME: 1 HOUR
SERVES: 6-8

120g unsalted butter
250g digestive biscuits
350g cream cheese, or vegan
  cream cheese
350g double cream, or vegan
  whipping cream
60g icing sugar
zest of 3 limes
3 kiwis, peeled and chopped

SWAPS

FLAVOUR
• No kiwis? Use pineapple,
berries, mangoes, plums.
• No limes? Use lemon zest.

Straight from the Oddbox hall of fame, this chilled-out cheesecake is the ultimate crowd-pleaser. It's mega easy to make too, and you can swap any fruit for the kiwis depending on what's in season and what you've got. If you have any leftovers, take a look at the cheesecake trifle below.

1. First make the biscuit base. Melt the butter in a small saucepan over medium heat. Put the digestive biscuits in a food processor and blend until you have crumbs. Place the crumbs in a bowl and pour over the melted butter. Mix well.
2. Transfer the base to a cake tin with a removable bottom. Push down the mixture into the bottom of the tin. Chill in the fridge for 30 minutes, until cold and firm.
3. Now make the cheesy bit. Put the cream cheese, cream, sugar and lime zest into a large mixing bowl. Use an electric whisk to mix until thick and smooth. Be patient - it will take a few minutes.
4. Spread the cheesecake mixture on top of the biscuit base and flatten the top. Arrange the chopped kiwis over the top.
5. Chill in the fridge for a minimum of 1 hour (or overnight) until it sets. Take out 20 minutes before serving.

LEFTOVERS?
STORE THE CHEESECAKE IN THE FRIDGE FOR UP TO THREE DAYS. AND IF YOU HAVE ANY LEFTOVER CREAM CHEESE MIXTURE, LAYER IT IN JARS WITH FRESH FRUIT, BISCUITS AND MELTED CHOCOLATE TO MAKE CHEESECAKE TRIFLES.

GRAPE CLAFOUTIS.

PREP TIME: 10 MINUTES
COOK TIME: 35 MINUTES
SERVES: 6

100g plain flour
250ml milk
6 eggs
pinch of salt
350g seedless grapes, stems
  removed
90g caster sugar

SWAPS

FLAVOUR
• No grapes? Use any fruit
- peaches, cherries, berries,
apples, apricots, pears.

This might have a fancy-sounding name, but it's one of the easiest desserts to make. It's the perfect way to use up an unloved-looking bunch of grapes in the fridge, or any other fruit in the bowl.

1. Preheat the oven to 180°C/160°C fan/350°F/gas 4.
2. Put the flour, milk, eggs and salt in a blender.
3. Blend to a smooth batter.
4. Pour the batter into a baking dish and add the grapes.
5. Bake for 35 minutes, then remove from the oven and immediately sprinkle the sugar on top while the clafoutis is still hot.
6. Leave it to cool, then serve.

LEFTOVERS?
KEEP YOUR CLAFOUTIS COVERED
IN THE FRIDGE FOR UP TO
THREE DAYS.

LEMON POLENTA CAKE.

PREP TIME: 15 MINUTES
COOK TIME: 40 MINUTES
SERVES: 6-8

170g unsalted butter, at room
   temperature
170g caster sugar
2 large eggs
170g ground almonds
85g polenta
zest and juice of 2 lemons

TO SERVE
Greek or vegan yoghurt
handful of berries (optional)

SWAPS

STRUCTURE
• No lemons? Use 1 orange,
2 mandarins, 2 limes, 2
satsumas, 2 clementines,
1 grapefruit.

This zingy gluten-free cake is made using polenta and ground almonds. We like it with a big dollop of Greek yoghurt and a scattering of fresh berries - or dial up the decadence by drizzling melted dark chocolate over the top. Fancy.

1. Preheat the oven to 180°C/160°C fan/350°F/gas 4, and line a cake tin or loaf tin with baking paper.
2. Place the softened butter and sugar in a large mixing bowl. Use an electric mixer to blend for 5-6 minutes, until pale and creamy.
3. Add the eggs one at a time, making sure the first egg is mixed in before adding the second one. Whisk with an electric mixer for 2-3 minutes.
4. In another bowl, combine the ground almonds and polenta. Add half the almond polenta mixture to the egg mixture and slowly stir with a spatula until combined. Add the rest of the almond polenta mix and stir until everything is well combined.
5. Add the lemon zest and juice and mix in.
6. Pour into the tin and bake for 40-45 minutes, until it's golden brown and a skewer in the middle comes out clean.
7. Let it cool in the tin, then serve with a dollop of yoghurt, fresh berries or the topping of your choice.

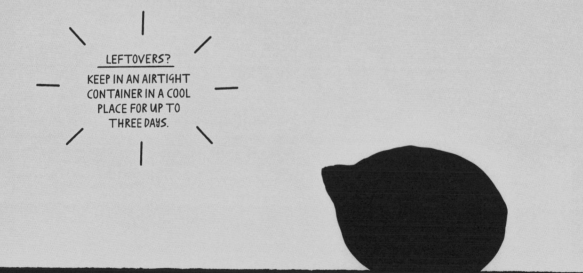

LEFTOVERS?
KEEP IN AN AIRTIGHT
CONTAINER IN A COOL
PLACE FOR UP TO
THREE DAYS.

PEACH TARTE TATIN.

PREP TIME: 15 MINUTES
COOK TIME: 35 MINUTES
SERVES: 6

8-9 peaches, depending on size
   (800g)
120g caster sugar
70g unsalted or vegan butter
1 tsp vanilla paste
1 packet ready-rolled puff
   pastry or vegan puff pastry

SWAPS

STRUCTURE
• No peaches? Use roughly
800g of plums, apples, apricots,
nectarines, pears, pineapple.

LEFTOVERS?

KEEP THE TARTE TATIN
IN THE FRIDGE FOR UP
TO THREE DAYS.

Looking for a showstopper that won't take an entire afternoon to make? With a simple vanilla caramel and juicy sliced peaches, you can't go wrong making this sticky-sweet dessert. There's no need to wait for it to cool down - serve warm from the oven with a scoop of ice cream or dollop of cream. Divine.

1. Preheat the oven to 180°C/160°C fan/350°F/gas 4.
2. Halve the peaches and remove the stones.
3. Put the sugar and butter in an ovenproof pan big enough to hold the peach halves in one layer, then place over a medium heat for 5-7 minutes or until you have a light golden caramel.
4. Add the vanilla paste to the caramel in the pan then carefully arrange the peaches, cut side up, on top.
5. Place the pastry on top of the peaches, and press the sides down gently to cover them.
6. Transfer the pan to the oven and bake for 30-35 minutes, or until the puff pastry turns golden brown and the caramel is bubbling.
7. Take the tart out of the oven and allow it to cool for a few minutes. Then tip it out on to a serving plate and enjoy warm or cold with ice cream or whipped cream.

PEAR & CHOCOLATE BROWNIES.

PREP TIME: 15 MINUTES
COOK TIME: 45 MINUTES
SERVES: 9

180g unsalted butter
190g dark chocolate, broken
 into pieces
3 eggs
250g golden caster sugar
100g plain flour
pinch of salt
20g cocoa powder
2 pears, cored and chopped
25g chopped nuts or seeds of
 your choice

SWAPS

FLAVOUR
• No pears? Use apples,
apricots, mangoes, pitted
cherries, raspberries,
strawberries.

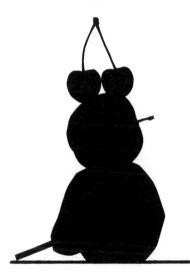

Chocolatey, fruity, nutty, yummy, gooey, crumbly, crunchy and chocolatey. Did we say fruity? Every bite of these brownies is a party of tastes and textures - and best of all, they're simple enough for even newbie bakers to try. Swap in any fruit you think goes well with chocolate, or try a different flavour every week. Why not?

1. Preheat the oven to 180°C/160°C fan/350°F/gas 4.
2. Fill a small saucepan with water to about a quarter of the way from the top. Pop a mixing bowl on top so it rests on the rim of the pan, but the bottom doesn't touch the water (this is a bain marie).
3. Put the butter and chocolate in the mixing bowl, and put the pan over a medium heat until the butter and chocolate have melted.
4. In a separate bowl whisk the eggs and sugar with an electric mixer for 4-5 minutes until they're thick and creamy.
5. Pour the chocolate into the egg mixture. Mix until combined.
6. Add the flour, salt and cocoa powder. Mix until combined.
7. Add the pears and nuts. Mix until combined.
8. Pour the batter into a lined 20cm square cake tin.
9. Bake for 40-45 minutes. Let the brownie cool completely before cutting into little squares and serving with a dollop of crème fraîche, a drizzle of custard or vanilla ice cream.

LEFTOVERS?
STORE IN AN AIRTIGHT
CONTAINER FOR UP TO
TWO WEEKS IN THE FRIDGE,
AND IN THE FREEZER FOR UP
TO ONE MONTH.

PANNA COTTA WITH POACHED FRUIT.

Panna cotta literally means 'cooked cream' – the perfect silky-smooth base for soft, stringy fruit. We've used rhubarb in the recipe because we love it. But if it's not in season or you're not a fan, you can use any other poach-able fruit, such as peaches, pears or plums. Camille also recommends swapping the vanilla for cardamom and using mangoes – how about that?

PREP TIME: 20 MINUTES
COOK TIME: 10 MINUTES
REST TIME: 3 HOURS
SERVES: 4

FOR THE PANNA COTTA
300ml any double cream
200ml any milk
1 tsp vanilla paste
30g caster sugar
2 gelatine leaves or 4g Sosa
  pro panna cotta powder (vegan
  gelling agent)

FOR THE POACHED RHUBARB
125g caster sugar
125ml water
2 rhubarb stalks (200g)

SWAPS

FLAVOUR
• No rhubarb? Use peaches, plums, pears, apricots, figs, mangoes.

LEFTOVERS?
STORE IN THE FRIDGE FOR UP TO THREE DAYS. SERVE ANY LEFTOVERS WITH CANTUCCINI OR AMARETTI BISCUITS CRUSHED ON TOP.

1. Put the cream, milk, vanilla paste and sugar in a saucepan over medium heat. Just before the liquid reaches boiling point, remove the pan from the heat.
2. Meanwhile, soften the gelatine leaves in ice-cold water for 3–4 minutes. Squeeze out the water and stir into the cream mixture until dissolved.
3. Transfer the mixture to a jug and pour into four glasses or upcycled jam jars, leaving a gap at the top.
4. Leave to cool completely, then chill in the fridge until fully set, for three hours or overnight.
5. To poach the rhubarb, place the sugar and water in a saucepan and bring to the boil. Cut the rhubarb into 5cm batons. Add the rhubarb to the boiling syrup, then immediately turn off the heat. Let the rhubarb poach and cook in the syrup until it's cold.
6. When ready to eat, use a slotted spoon to remove the rhubarb from the syrup and serve on top of the panna cotta.

SURPRISE SPICED CRUMBLE.

PREP TIME: 15 MINUTES
COOK TIME: 50 MINUTES
SERVES: 6

8 plums (750g)
1 tsp vanilla paste (optional)
200g plain flour
80g ground almonds
120g light muscovado sugar
1 tsp mixed spice
1 tsp ground cinnamon
1 tsp ground ginger
220g cold butter or vegan
    butter (not vegan margarine)
pinch of salt

SWAPS

FLAVOUR

• No plums? Use peaches, apples, pears, blueberries, blackberries, pineapple, mangoes, apricots. Or try these pairings: blueberries and apples, blackberries and pears, pears and apples, plums and blueberries, apricots and apples.

LEFTOVERS?

KEEP IN AN AIRTIGHT CONTAINER IN THE FRIDGE FOR UP TO THREE DAYS. EAT COLD OR REHEAT IN THE OVEN FOR ABOUT 10 MINUTES, OR UNTIL WARM.

Having a fail-safe buttery topping in your repertoire means you can whip up a crumble with any fruit in the bowl - even a pineapple. You can use more than one fruit if you fancy (see ideas below). If you don't have ground almonds, leave them out and add the same amount of extra flour instead. And don't be afraid to mix up the spices with nutmeg or cardamom. Surprise!

1. Preheat the oven to 180°C/160°C fan/350°F/gas 4. Halve the plums and put the stones in the compost.
2. Cut the plum halves in half again and place in a mixing bowl with the vanilla paste and a tablespoon of water. Toss to coat the pieces and place in a baking or pie dish.
3. In a large bowl, mix all the dry ingredients together: flour, ground almonds, sugar and spices.
4. Cube the butter and add it to the flour mixture. Rub together with your fingertips until it looks like breadcrumbs. (Don't worry if the crumbs are different sizes - they'll give your crumble a lovely texture.)
5. Scatter the crumble over the fruit.
6. Bake for about 45-50 minutes until the topping is deep golden brown and the fruit is bubbling through the crust.
7. Serve warm from the oven, or at room temperature.

WHOLE ORANGE & ALMOND CAKE.

PREP TIME: 1 HOUR
COOK TIME: 35 MINUTES
SERVES: 8

2 oranges (450g)
1 tbsp olive oil, for greasing
6 eggs
190g caster sugar
1 tsp vanilla extract
250g ground almonds
1 tsp baking powder

SWAPS

STRUCTURE
• No oranges? Use 450g
clementines, mandarins or
satsumas. (Don't use grapefruit
- their white membrane is too
bitter.)

LEFTOVERS?

TO MAKE CAKE TRUFFLES:
CUT YOUR LEFTOVER CAKE
INTO SMALL PIECES, LAY
THEM ON A BAKING RACK
OVER A BAKING TRAY AND
POUR MELTED CHOCOLATE
OVER THEM. LEAVE THEM
TO SET, THEN CHILL IN THE
FRIDGE. IF YOU MAKE A BIG
BATCH, YOU CAN FREEZE
THEM FOR MONTHS –
THE PERFECT SNACKABLE
SWEET TREAT.

Camille says: 'This zero-waste cake recipe is super-easy to make, deliciously fluffy and can be made using gluten-free baking powder. Feel free to add a few chocolate chips to the batter for a classic orange and chocolate flavour pairing. And have a look at the leftover tips for my favourite chocolate-cake truffle recipe.'

1. Zest one of your oranges into a bowl and set aside to sprinkle over the cake later. Then place all the oranges in a saucepan and cover with water. Cover with a lid and bring to the boil.
2. When the water boils, reduce the heat to a simmer and cook for 50-60 minutes until the oranges are completely soft. (Or cover the oranges in water, place them in the microwave and cook for 10 minutes.)
3. Meanwhile, grease the sides and bottom of a 30cm loose-bottomed cake tin (or two smaller ones).
4. Remove the oranges from the water. Set aside to cool.
5. Preheat the oven to 180°C/160°C fan/350°F/gas 4.
6. When they're cool enough to handle, cut the oranges into quarters and remove any seeds. Place the oranges - skin on - in a food processor and blend to a fine purée.
7. Place the eggs and sugar in a large mixing bowl. Use an electric mixer to whisk them together until white and fluffy - about 3-4 minutes. Then whisk in the orange, vanilla, ground almonds and baking powder.
8. Pour the orange batter into the greased tin and bake for 30-35 minutes, or until an inserted skewer comes out clean. Set aside to cool slightly, then remove from the tin, sprinkle the orange zest on top and serve with a dollop of cream.

FRUITY FRITTERS.

PREP TIME: 10 MINUTES
COOK TIME: 10 MINUTES
SERVES: 2

100g self-raising flour
1 egg
70ml milk
2 apples (200g)
2 tsp ground cinnamon
4 tsp caster sugar
1 tbsp butter, for frying

FOR VEGAN BATTER
120g plain flour
1 tsp baking powder
75ml plant-based milk
75ml water
1 tbsp vegan butter or
  vegetable oil

SWAPS

STRUCTURE
• No apples? Use pears,
peaches, pineapple or bananas
cut lengthways.

Apples in the autumn, peaches in the summer. These fritters are so versatile you can use almost any fresh fruit you have at home. What's more, they only take 20 minutes to make – flash fritters in every sense.

1. Put the flour, egg and milk into a mixing bowl. Mix until combined and put to one side. (For the vegan batter mix all the ingredients together and set aside.)
2. Core and slice the apples into 1cm thick rings.
3. Put the apple rings into the batter and mix until well coated.
4. In another mixing bowl, combine the ground cinnamon and sugar. Set aside.
5. Melt the butter in a large frying pan over a medium-high heat. (For vegan batter use vegan butter or oil.)
6. Lift one apple ring at a time from the batter and gently place in the pan. Cook the fritters for 3-4 minutes on each side, until puffed and golden. Repeat until you've cooked all the apple rings.
7. Toss the fritters in the cinnamon mixture until coated and tuck in.

LEFTOVERS?

KEEP ANY EXTRA FRITTERS IN A CONTAINER IN THE FRIDGE FOR UP TO THREE DAYS OR IN THE FREEZER FOR THREE MONTHS. THEY CAN BE PUT BACK INTO A HOT OVEN TO CRISP UP IF NEEDED.

BAKED RICE PUDDING.

PREP TIME: 10 MINUTES
COOK TIME: 1 HOUR
30 MINUTES
SERVES: 6

2 tbsp unsalted or vegan butter,
  for greasing
2 apples, grated (250g)
120g pudding rice
40g honey or maple syrup
1 tsp ground nutmeg
1 tsp vanilla extract
30g raisins (optional)
800ml whole milk or plant-
  based milk

SWAPS

FLAVOUR
• No apples? Use grated pears,
chopped peaches, apricots,
plums, bananas.

Remember rice pudding from school? You do? Well, luckily this is nothing like that. We've used grated apple in the recipe, but feel free to go wild with other chopped fruit. And for extra zing add a sprinkling of orange zest too. Seconds, please!

1. Preheat the oven to 160°C/140°C fan/320°F/gas 3 and grease the baking dish.
2. In a bowl, mix the grated apples, pudding rice, honey, nutmeg, vanilla, raisins and milk.
3. Pour into the buttered dish and bake for 1 hour 30 minutes, or until the rice is tender and the top is golden and just set.
4. Scoop the warm pudding into bowls and serve with fresh fruit, ice cream or a dollop of fruit compote.

LEFTOVERS?
KEEP IN THE FRIDGE FOR
UP TO THREE DAYS.

COOKING SCRAPPY.

TURN SEEDS, SKINS AND PEEL INTO SOMETHING DELICIOUS.

188

182

CRACKING CRISPS.

PREP TIME: 5 MINUTES
COOK TIME: 15 MINUTES

vegetable and fruit peelings
 - potato, carrot, beetroot,
 parsnip, apple
olive oil
salt

LEFTOVERS?

THE CRISPS WILL
KEEP FOR 1–2 DAYS
BUT ARE BEST EATEN
STRAIGHT AWAY.

Peeled your spuds, roots and carrots? Hold on to the skins to make your own crisps. You can do this while you cook or build a collection to make later. Top tip: have a go at the hummus on page 138 and the Smoky veg dip on page 146 for the ultimate no-waste snack.

1. Preheat the oven to 200°C/180°C fan/400°F/gas 6.
2. In a bowl, mix the peels with a little oil and season with salt.
3. Lay the oiled peels on a baking tray, separating each one so they crisp up.
4. Pop them in the oven for 5–15 minutes, depending on how thick they are.
5. Give them a toss about halfway through cooking, then cook until nice and crispy.

HERBY BUTTER CUBES.

PREP TIME: 15 MINUTES

100g any butter, softened, or oil
small bunch of parsley or any
  other soft herb, finely chopped
1 small red chilli, finely chopped
1 garlic clove, grated
½ tsp ground nutmeg (optional)
pinch of salt

We've all been there: the recipe calls for a handful of herbs and you have a whole bag of the stuff. Instead of wasting what's left over, try making these butter cubes to add an easy herby hit to future dishes. Use them to sauté veg instead of oil or butter, or add them at the end of a recipe for extra flavour.

1. Pop the butter into a bowl then mix through the parsley, chilli and garlic.
2. Mix in the nutmeg and a good pinch of salt.
3. Either wrap the butter in greaseproof paper and store it in the fridge, or scoop individual blobs into an ice cube tray to freeze.

LEFTOVERS?
KEEP IN THE FRIDGE FOR ONE
WEEK, OR IN THE FREEZER
IN INDIVIDUAL CUBES FOR
THREE MONTHS.

SCRAPPY STOCK.

PREP TIME: 5 MINUTES
COOK TIME: 30 MINUTES
SERVES: 2

vegetable scraps, including
    skins and peelings
1 bay leaf (optional)
black peppercorns (optional)

LEFTOVERS?

THE STOCK CAN BE
STORED IN THE FRIDGE
FOR UP TO FIVE DAYS OR
IN THE FREEZER FOR
THREE MONTHS.

Homemade stock tastes better than ready-made. Fact. And stock that helps you use up every scrap? Well that tastes the best. You can chuck in all your odds and ends, but avoid beetroot, starchy veg (such as potatoes and butternut squash) and bitter greens (like kale, broccoli and cabbage). And if you're mainly using skins and peel, add some fresh veg, too, or it won't have much flavour.

1. Pop the veg scraps into a pan with the bay leaves and peppercorns, if using, and add water to cover everything generously.
2. Bring to the boil, then reduce to a simmer for 30–40 minutes. Strain and compost the scraps.
3. Rapidly boil the stock to reduce the volume and intensify the flavour.

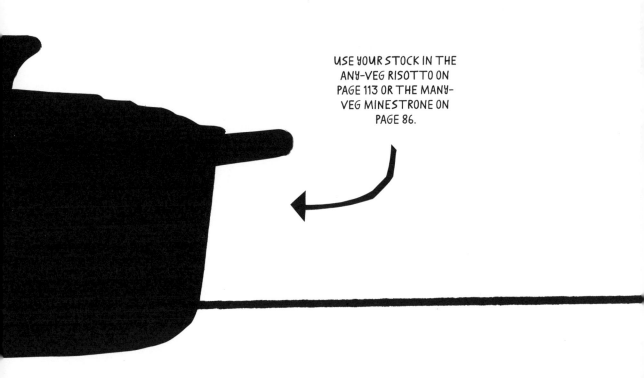

USE YOUR STOCK IN THE
ANY-VEG RISOTTO ON
PAGE 113 OR THE MANY-
VEG MINESTRONE ON
PAGE 86.

ROASTED & TOASTED SEEDS.

PREP TIME: 10 MINUTES
COOK TIME: 15 MINUTES

200g raw pumpkin or squash
  seeds
20ml olive oil
pinch of salt

LEFTOVERS?

THE SEEDS WILL KEEP
FOR MONTHS IN AN
AIRTIGHT CONTAINER.
OR YOU CAN USE THEM
IN THE SEEDY DUKKAH
OPPOSITE.

Snacks, toppings, sprinkles and more. There's loads you can do with pumpkin and squash seeds - gather them up as you cook and then roast a big batch in one go.

1. Preheat the oven to 180°C/160°C fan/350°F/gas 4.
2. Pop the seeds into a bowl of water and give them a good rinse. When they float to the top, scoop them out and put them in a saucepan.
3. Cover the seeds with water, add a pinch of salt and simmer for 5 minutes. Drain and spread out on a clean tea towel then pat dry.
4. Spread the seeds on a baking tray, drizzle over a little oil and sprinkle with salt and any other spices you fancy.
5. Roast in the oven for 10-15 minutes or until golden brown.

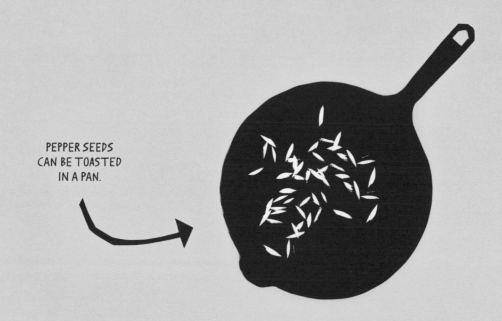

PEPPER SEEDS
CAN BE TOASTED
IN A PAN.

SEEDY DUKKAH.

PREP TIME: 15 MINUTES
COOK TIME: 20 MINUTES

50g toasted butternut
  squash or pumpkin seeds
  (see opposite)
50g walnuts, or other nuts
3 tsp cumin seeds
3 tsp coriander seeds
3 tsp fennel seeds
30g sesame seeds
pinch of salt
sprinkling of ground cinnamon
  (optional)

Dip, dab, repeat. Dukkah is an Egyptian condiment made up of nuts, herbs and spices - we've given it a twist by adding roasted and toasted seeds that you might otherwise have thrown away. Sprinkle over salads, potatoes, veg for roasting, couscous or rice dishes, or use as a dip with chunky bread and oil.

1. Preheat the oven to 180°C/160°C fan/350°F/gas 4.
2. Put the toasted seeds in a bowl.
3. Spread out the walnuts on a tray and roast in the oven for 5-10 minutes or until nicely toasted. Allow them to cool then crush with your hands into the bowl.
4. Meanwhile, gently toast the cumin, coriander and fennel seeds in a frying pan until they start to pop. Give them a quick pulse in the blender, then add to the bowl with the walnuts and seeds.
5. Toast the sesame seeds in the frying pan until nice and golden, then add to the bowl and give everything a good mix.
6. Add a good pinch of salt and a sprinkling of cinnamon, if you like.

LEFTOVERS?
KEEP IN AN AIRTIGHT CONTAINER IN THE CUPBOARD FOR A MONTH OR SO.

LEFTOVER LIMONCELLO.

PREP TIME: 10 MINUTES
COOK TIME: 5 MINUTES
REST TIME: 2 WEEKS

24 squeezed lemon halves
1 litre good-quality vodka
700ml water
700g sugar

Planning a dinner party? This traditional Italian liqueur is a great one to make in advance – it takes a bit of prep, but the results are totally worth it. Instead of squeezing lemons specially, keep the leftover halves from other recipes in a bag in the freezer until you have enough.

1. Sterilise a big, airtight jar in a pan of boiling water (see page 31).
2. Put the lemon skins and vodka into the jar. Make sure the vodka covers the lemons – push a circle of parchment paper into the jar to stop them bobbing above the surface if needed.
3. Seal the top and leave for one week at room temperature, shaking the jar every day.
4. A week later, make a sugar syrup by bringing the water and sugar to the boil in a saucepan until the sugar dissolves.
5. Combine the sugar syrup and the lemon liquid and leave for another week, shaking the jar every day.
6. Strain the liquid and compost the lemons. Fill sterilised, clean bottles and enjoy (in small quantities!).

<u>LEFTOVERS?</u>

THE LIMONCELLO WILL KEEP FOR MONTHS – YOU CAN EVEN KEEP IT IN THE FREEZER TO GIVE IT A LOVELY SYRUPY CONSISTENCY.

PICKLED STALKS & STEMS.

PREP TIME: 5 MINUTES
COOK TIME: 15 MINUTES

150ml vinegar
50ml water
½ tsp caraway seeds
½ tsp coriander seeds
20ml hot sauce (optional)
1 broccoli or cauliflower stem
1 small beetroot (100g)

Why should florets have all the fun? The stalks and stems of broccoli and cauliflower are just as nutritious - have a go at this super-simple pickling recipe and add a whack of flavour to salads, curries, toasties and burgers.

1. Put the vinegar, water, caraway seeds, coriander seeds and hot sauce, if using, into a pan and bring to a simmer.
2. Cut the broccoli or cauliflower stem and the beetroot into thin matchsticks or small cubes. (No need to peel anything unless you want to.)
3. Turn the heat off under the pan then carefully add the veg and mix them around. Leave for 5–15 minutes, stirring every so often. The veg will still be slightly crunchy but will have taken on a subtle warmth and tang.
4. Pop into a jar or container once cool.

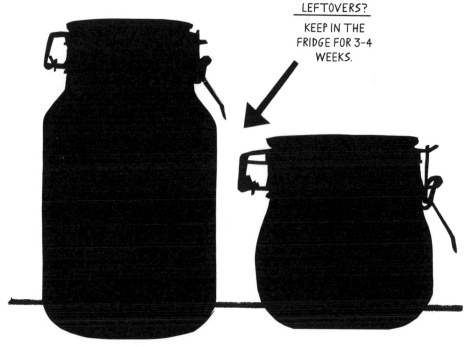

LEFTOVERS?

KEEP IN THE FRIDGE FOR 3-4 WEEKS.

ODDS & ENDS SAUCE.

PREP TIME: 10 MINUTES
COOK TIME: 30 MINUTES

50ml olive oil
300g veg odds and ends,
  roughly chopped
4 garlic cloves, sliced
2 tins tomatoes (800g)

LEFTOVERS?

THE SAUCE WILL
KEEP IN THE FRIDGE
FOR UP TO THREE
DAYS OR IN THE
FREEZER FOR THREE
MONTHS.

Carrot tops, spring onion tips and celery leaves at the ready - this simple sauce is here to keep food bins empty and plates full. Store all your odds and ends in a bag in the freezer, ready to cook. Then make this sauce to have with pasta or use it as a base for lasagne, stews and curries.

1. Heat the oil in a big saucepan and add your odds and ends. Fry for a few minutes until starting to turn soft.
2. Add the sliced garlic with the tinned tomatoes, then put a splash of water into the tins, swirl it round to catch the last drops of tomato juice and add to the pan.
3. Bring the sauce to a simmer and let it bubble away for 30-40 minutes.
4. Blitz to a smooth sauce in the blender or keep it chunky.

USE YOUR SAUCE
IN THE NO-LIMITS
LASAGNE ON PAGE 100.

VEG-PEEL POWDER.

PREP TIME: 5 MINUTES
COOK TIME: 5-6 HOURS

carrot peelings
onion and garlic skins
sea salt
chilli flakes

About to chuck your peel and skins? Hold on to them to make these super-simple powders – combined with salt and chilli flakes, they're a great way to add a sprinkling of flavour to your cooking.

1. Lay the peel and skins on a baking tray and pop them into the oven on the lowest setting for 5–6 hours, until bone dry and crispy (or use a dehydrator if you have one).
2. Set to one side and allow to cool.
3. Use a blender to blitz them into a powder.
4. Add salt and chilli flakes to create your own seasoning.

LEFTOVERS?

THE POWDERS CAN BE
STORED FOR MONTHS
IN AN AIRTIGHT
CONTAINER.

FRUIT-SCRAP VINEGAR.

PREP TIME: 5 MINUTES
REST TIME: 4 MONTHS

2l water
200g sugar
500g fruit scraps
100ml live vinegar, or any
   good-quality vinegar

Don't be put off by the resting time – this vinegar is really easy to make and is a great way to use up all your fruit scraps and peel. Throw in apple cores, melon skin, mango peel and whatever else you have. Once the vinegar's ready, use it in dressings and to give your dishes a tangy kick.

1. Sterilise a three-litre airtight jar in a big pan of boiling water (see page 31).
2. Put all the ingredients into the jar – you want the fruit to take up about three-quarters of the space. Give it a good mix, then loosely cover it with cheesecloth or tissue paper. (This keeps out any bugs but allows the ingredients to ferment.)
3. Leave the jar on a shelf for two weeks, stirring the contents once a day. It will start to look milky – this is a good thing.
4. After two weeks, strain the liquid and pour it into a clean, airtight jar or bottle. Lock the top and forget about it for a couple of months.
5. Taste the vinegar every so often to check the acidity level – use it when you're happy with the taste.

LEFTOVERS?
VINEGAR GETS BETTER WITH AGE, SO
TAKE YOUR TIME TO USE IT. AND IF
YOU HAVE THE SPACE, KEEP MAKING
MORE TO BUILD A COLLECTION.

INDEX.

HarperCollins*Publishers*
1 London Bridge Street
London SE1 9GF

www.harpercollins.co.uk

HarperCollins*Publishers*
1st Floor, Watermarque Building, Ringsend Road
Dublin 4, Ireland

First published by HarperCollins*Publishers* 2022

10 9 8 7 6 5 4 3 2 1

A catalogue record of this book is available from the British Library

ISBN 978-0-00-855448-4

Photography by Ola Smit
Food styling by Esther Clark
Prop styling by Anna Wilkins
Recipes by Martyn Odell and Camille Aubert
Copywriting by Maia Swift
Produced by Alice Ratcliffe
Illustration by Alex Green

Printed and bound in Bosnia and Herzegovina by GPS Group

WHEN USING KITCHEN APPLIANCES PLEASE ALWAYS FOLLOW THE
MANUFACTURER'S INSTRUCTIONS

A PUMPKIN-SIZED THANK YOU TO...

Martyn Odell for working with us to create the concept
and recipes. Camille Aubert for contributing and testing
recipes. Maia Swift for writing the book. Alice Ratcliffe
for running the show. Alex Green for his design and
illustration wizardry. Ana Caeiro, Heather Lynch and
Sam Boggis-Rolfe for their ideas. Emilie Vanpoperinghe
and Deepak Ravindran for trusting us to get on with it.
The team at HarperCollins for making it real.
Our growers for inspiring us with their rescue stories
every week. The wonderful Oddbox community
- we wouldn't be here without you.